Indisputable Proof

That CO_2 Does Not Dominate Global Climate

Indisputable Proof

That CO_2 Does Not Dominate Global Climate:

GEO-PHYSICAL HISTORY OF EARTH, CARBON DIOXIDE AND
EXTRATERRESTRIAL FORCES AND GLOBAL CLIMATE

Bobby E. Leonard, Ph.D.

&

Arthur C. Lucas, Sc.D.

International Academy of Hi-Tech Services, Inc.

1007 Rosslare Court, Arnold, Maryland 21012, USA.

Published by International Academy of Hi-Tech Services, Inc.
Arnold, MD

FIRST EDITION

Indisputable Proof, That CO2 Does Not Dominate Global Climate/
Bobby E. Leonard & Arthur C. Lucas
ISBN-13: 978-1511926607
ISBN-10: 1511926600

Climate, Climate Change, Global Warming, Earth History

Dr. Leonard offers his overwhelming gratitude to his dear wife, Georgia Carol Beecher, to endure for him the many hours, days, weeks and months during our preparation of this manuscript. Dr. Leonard expresses his appreciation to Dr. Richard McElroy for his editing and suggestions on manuscript layout.

Dr. Lucas expresses his appreciation for his wife's help in editing the manuscript.

TABLE OF CONTENTS

LIST OF TABLES

LIST OF FIGURES

PREFACE

Much of what has been presented in this manuscript is common knowledge to not only geophysicists but to the lay person that view televisions excellent educational channels such as the National Geographic and Discovery channels. In many cases, there are more than one theory (or even many) such as the origin and details of the Moon becoming part of Earth's system around the Sun. In these cases, we have presented what we believe to be the most probable scenarios and present them as "probable". There are many proxies with respect to Earth's past climate, present status and supportive evidence for Earth's climate future. We have relied heavily on the ice core data because we believe they provide the most accurate and consistent for our past climate. As we pointed out, the ice core data for CO_2 are technically not proxy data (i.e. representative agent for CO_2) but, as shown in the Figure 55, tiny bubbles provides direct samples of CO_2 from the ancient past. There are three primary sources of ice core uncertainties, timescale, diffusion, and sampling. Altogether, these are estimated to be resolvable to within $\pm 3\%$ standard deviation (Steig 2005). We are re-assured by NOAA's extensive support of the many American and world-wide groups engaged in ice core data collection, analysis and journal support for publication. Our work is as a result of work by others such as Siegenthaler et al (2005), Petit et al (1999), Siddall et al (2003), Soon et al (2007), Pagani et al (2005, 2006), Indermuhle et al (2000), Dahl-Jensen et al (2012) and many, many more. With complete confidence in the Antarctica, Greenland, deep ocean and other data, we show indisputable evidence that airborne carbon dioxide concentrations has not in the past, does not now dominate our Global Climate and in particular does not dominate Global Surface Air Temperature. Further, that all periods of the Glacial-to-Interglacial- back to-Glacial global climate behavior is dominated by extraterrestrial forces, and indisputably the current Holocene and past Interglacial Periods. In Figures 44 and 45, we show that most probably the three astronomical forces, even in the interglacial periods vary significantly and are far from being understood. Thus, we must accept that we do not understand the mechanisms of our past, present and future Global Climate. But this work does prove that the astronomical forces dominate our global climate and carbon dioxide plays only a small secondary role.

Chapter 1 — Introduction

This book is intended to trace the history of Earth and in particular highlight the significant events that have occurred that have made Earth as we see it today. This includes the factors that have been favorable and the factors that have been unfavorable. Even though we are in danger of over-populating the Earth, it is obvious that humans have had, up to this stage of Earth's evolution, minimal effect on the way Earth geo-physically manages itself today. Humans even now have no means of controlling most of the events that shape Earth's destiny such as volcanic eruptions, earthquakes, tornadoes, hurricanes, tsunamis, plate tectonics movement or asteroid impacts. We are in some cases able to predict these events before they happen but not prevent them. We can take actions that can minimize the impact of the events such as constructing stronger, earthquake resistant buildings (i.e. San Francisco), building stronger dikes against hurricane sea surges (i.e. New Orleans), National Aeronautics and Space Administration (NASA) aerial and satellite hurricane tracking systems, establishing ocean warning systems for tsunamis, and setting up international watch systems for potentially dangerous asteroids. With the advent of the atomic age by the development of the atomic bomb, mankind however now has the capability of delivering significant harm to life on Earth including our homo-sapiens species. Humans do have the capability of drastic alteration of our atmosphere and hydrosphere. If mankind becomes extinct by his own doings, we will leave behind world-wide sustainable landmarks of asphalt and concrete as portrayed in the movie "Planet of the Apes". Due to energy demands, Mankind some say has also entered into an evolutionary phase where his environment and climate may be in jeopardy with uncontrolled carbon dioxide emissions. As we progressed through Earth's time-table, we have paused at points where aspects are uncertain such as – how we got our Moon? So this book, with the initial primary theme being to provide a comprehensive history of Earth (as to how we got here and what future is expected to bring), instead devotes considerable space in the later sections to Global Warming because it appears to some, including we the authors, to represent mankind's largest environmental and economic challenges. A significant controversy exists relative to the influence of the greenhouse gases (GHG), in particular airborne carbon dioxide from human activities, on the global climate. The reports by the international Intergovernmental Panel on Climate Change [IPCC, 1990, 2007] estimates that, in response to a doubling of pre-industrial airborne carbon dioxide (CO_2) concentration, the mean global surface air temperature will rise 3.26 °Kelvin (1 °Kelvin = 1 °K = -273 .15°Centagrade – absolute zero. A change of 1°Kelvin is a change of 1°Centegrade). Although life on earth has thrived at much, much higher levels of carbon dioxide, the most recent White House study predicting detailed deleterious effects of human–made global warming and the Congressional drafting of the Cap and Trade global warming bill emphasizes global warming concerns. The White House report states "Warming of the climate is unequivocal, and man-made gases are primary to blame." in spite of the large contingent of geo-scientist that think otherwise. Recent revelation that some of the data supporting human GHG production as a significant source causing increased global temperatures was fraudulently fabricated and has put the global warming theory into serious jeopardy, at least in the minds of many scientists and much of the general public. Irrespective technically, there are still two basic issues relative to the controversy. IPCC (1990, 2007, 2013a) estimates that atmospheric carbon dioxide is by far the dominant cause of Global Warming. Foremost of importance is the unknown correct correlation between atmospheric carbon dioxide variation and global surface air temperature. Second is the related question of whether, in the past large fluctuations in global CO_2 and global temperature, are the fluctuations driven by change in temperature first or change in CO_2

first, i.e. which leads and which lags? This is fundamental to the first question. Probably the most significant feature of this book then is the presentation, in the later chapters, of new data and analysis showing that Global Climate change is indeed not primarily driven by atmospheric carbon dioxide concentrations but rather by extraterrestrial forces affecting Earth's radiant energy flux at the top of our atmosphere. There are as many theories about the feedback effects and interpretation of climate data as there are geophysicists. This book however does not intend to venture into the evaluating all the feedbacks relative to the Global Climate Change controversy, but do simply wish to provide what we consider as an unique analysis of what is considered to be the most reliable past climate, carbon dioxide and temperature data published primarily by United States government agencies. A significant portion of the analysis is from the ice core data obtained from the Antarctica and Greenland ice caps by US and European research groups and published in National Ocean and Atmospheric Administration (NOAA) reports.

The opportunity presented here affords us to explore history of how our Earth has evolved into what can only be termed a remarkably beautiful place in our universe. In going back to the Big Bang, we complete the evolution process of how our galaxy came to be, our Solar system came to be and how our Earth and we human beings came to be. So it is appropriate first to go all the way back to the Big Bang creation of our universe and everything around us as we know it today. What we will show is some truly remarkable events in the Earth's evolution. To be technical, we humans are indeed a part of the initial Big Bang products in that the exact same hydrogen atoms that came out of the Big Bang are scattered throughout our universe and the exact same are even in our own bodies, which is 99% Big Bang hydrogen (13.82 Billion years old). All hydrogen are direct progeny of the Big Bang and all other elements on Earth in our Solar system and our universe came from the fusion of hydrogen. The ice age data from the ice cores show that it is most probably inevitable that we will most likely experience a further warming spell and then a severely cold ice age climate, again dropping the Global Surface Air Temperature as much as 13 °C (23 °Fahrenheit) as happened in the past. Earth having experienced eight such severe events in the past million years the odds are strongly in favor. However, as good news, in the later chapters, a reasonably accurate assessment is made of what our climate will be like for the next 10,000 years giving us many generations in time to prepare for this ice age. It is strongly believed that a true understanding of what to expect in the future here on Earth can only be made by understanding what Earth has been through in the past. This then identifies where we are, how and why we got here and what our future can be expected to be. The early chapters are thus devoted to how the universe has provided our Earth and how Earth's climate has progressed to its present state – a nearly ideal climate for Earth's flora, fauna and in particular we humans. It is shown that the Earth escaped near disasters in a number of instances in the past, whereas our nearest neighbors Venus and Mars have not fared so well in terms of providing conditions for life as we know it. It is shown that all three started out essentially the same but due to slightly different conditions, became absolute extremes. Much of what is provided here regarding this evolutional process is common knowledge, but it is hoped that by putting it all together in its proper sequence with documentation that is accepted by the world's prominent scientists and even Vice President Al Gore. He even acknowledges, in his text "An Inconvenient Truth" (Gore 2006), the ice core documentation of the Younger Dryas ice age reversal cold event cited on page 120, was not influenced by CO_2 forcing and primarily documented with the ice core data. For the more inquisitive, we in greater detail aim to let you readers see that our present climate is fortunately at its optimum and our climate is destined to undergo extreme changes in the future but hopefully, as stated above, with enough time for our technology and human adaptability to be able to cope with such extremes as seen in the past i.e. nearly all of Canada, now populated with 60 million people, was covered with kilo-meter thick sheets of ice. We devote a section of this text relative to the extreme hardships and drain on our energy resources when the next ice age does come. What we show is that we most probably will experience significant Global Warming as was experienced about 430,000 years before present (ka BP), primarily from extraterrestrial Milankovitch (1998) like forcing. This may last for another 20,000 years. The truly world-wide crisis will follow, the next Ice Age, but we will have time to prepare.

This book is divided into Parts, grouping Chapters with a common subject. Each Part is first summarized in the beginning. In Part I, we show where our Solar System and hence Earth is located in our Milky Way galaxy. We thus cite distances and velocities with respect to the locations of relative stellar objects. For those inclined, we provide three Appendices providing more extensive formulations to supplement the main text. In Appendix A, we present the methods for determining these astronomical distances and velocities. For distances, we provide the formulas for the Parallax method and, for velocities, we present the light spectra Red Shift Doppler method. In Appendix B, we provide the equations for the calculation of the Giant Impact Object's approach and eventual collision with Earth, this resulting in the formation of the Moon. Appendix C provides the equations to compute the correlation and covariance examining the similarity and relationship between the 430 ka BP interglacial warm period and our present Holocene Interglacial period.

PART I – FROM THE BIG BANG TO WHERE WE ARE IN OUR UNIVERSE

In this Part I containing Chapters 2 through 5, we provide for the formation of the universe as we know it today based on what our astrophysicists have learned, most recently through advanced radiofrequency, X-ray and visual telescopic devices such as the Hubble Telescope. Astronomers are able to see what has transpired in the past by viewing great distances in space. For example the Hubble telescope can see cosmos space to a distance of 13.2 billion light years (1 light year = 9.46 trillion kilometers) and hence what was happening 13.2 billion years ago in our universe (just about 0.6 billion years after the Big Bang).

Chapter 2 – The Beginning

In this chapter we will trace the evolution of our universe, our galaxy and our Solar system. From observation of innumerable stellar systems in our local galaxy, the Milky Way, we know that our Solar system's Sun is at least a second generation product of a first generation star that underwent a supernova explosion about 5 billion years ago. I use the term first generation because we know that stars form by gravitational attraction of cosmic matter, but all have a limited evolutionary time before reaching a number of possible fates depending on their size. Star formation happened only a few million years after the Big Bang (BB). We can go backwards even further in time to the creation of our universe, hypothesized by the Big Bang Theory (BBT), about 13.7 billion years ago. The origin of our universe and the BBT is not proven but it is known that our universe indeed did begin and began in an amazingly small space, about the size of Earth, and soon after the Big Bang consisted of only protons and electrons that were accelerated outward from this small space (most scientist indeed concede there was initially a "Big Bang"). The first defined stages of the universe after the Big Bang are the Inflationary and Nucleosynthesis stages. Nucleosynthesis refers to the production of nuclei other than those of the lightest isotope of hydrogen during the early phases of the universe. There still remains the question of what occurred before the Big Bang? Theologians have their own hypotheses (i.e. Spitzer 2010). High energy physics offers several theories, however the simplest is that the pre-Big Bang singularity occurred from the gravitational contraction of a previous universe, as is hypothesized as one fate of our present universe, called the Big Crunch Theory, perhaps to occur again, occurring +100 billion years in the future. Recently the fundamental Higgs Boson (particle) has been verified, which has been termed the "God particle" because it is the most fundamental with respect to the Big Bang and could even be the source of a mysterious Dark Matter and Energy which now is known to permeate significant space in our universe. (Peter Higgs, a Scottish physicists, and Fancois Englert, a Belgium physicist, just won the 2013 Nobel Prize for physics for their work relative to the Higgs particle.) All this is often proposed as part of an oscillatory universe scenario. A strong conformation of the BBT was the image recently obtained from the Wilkinson Microwave Anisotropy Probe of the full sky (Bennett et al 2003) shown as Figure1.

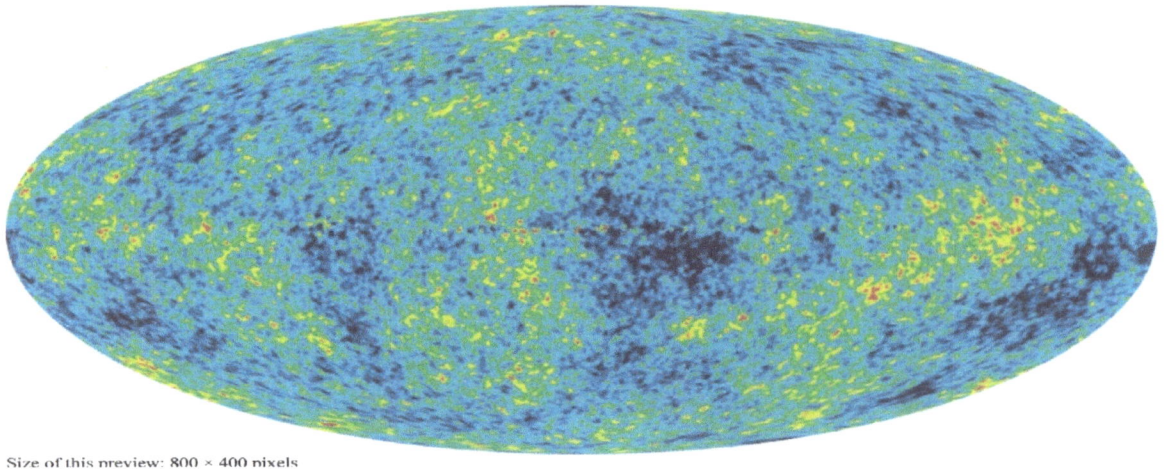

Size of this preview: 800 × 400 pixels

The Figure 1 shows the degraded residual background radiation from the Big Bang event. The average temperature is 2.725 ° Kelvin (degrees above absolute zero – absolute zero is the temperature where there is no molecular motion = - 273.16°Celsius (Centigrade). The microwave radiation and its image was first discovered in 1964 by radio astronomers Arno Penzias and Robert Wilkson, who received the Nobel Prize for their work. The colors represent the tiny temperature fluctuations, as in a weather map. Red and yellow regions are warmer and blue regions are cooler by about 0.0002 degrees. This image was acquired on 29 April 2008. Hydrogen and helium atoms began to be formed about 3 minutes after the Big Bang. Initially they were ionized, but as the universe cooled, the electrons were captured causing the emission of X-rays, a period lasting until about 310,000 years after the Big Bang. Astronomers are still searching for the initial X-ray radiation from the neutralization of the hydrogen and helium atoms by free electrons (Helium was formed from the fusion of hydrogen.). High energy physics encompasses elementary particles, originally the Higgs bosons, the formation of quarks at 10^{-6} seconds after the Big Bang and then the binding of the quarks to form hadrons such as protons and neutrons in an intense environment of photons.

From about 3 minutes to 20 minutes after BB, Nucleosynthesis occurred with the protons and neutrons combining into nuclei by nuclear fusion, forming deuterium (heavy hydrogen with a neutron – 2H), the helium isotopes 3He and 4He, the lithium isotopes 6Li and 7Li and the beryllium isotopes. No higher massed isotopes are believed to have been formed during this 17 minute Nucleosynthesis period. The mass percentages were 75% (92% by atom) hydrogen, 25% (8% by atom) helium and about 0.01% other elements. After this brief Nucleosynthesis period, the early universe cooled to non-fusion temperatures leaving a quasi-stable expanding cloud of primarily hydrogen and helium and photons. For this brief period then the universe was relatively dark with radiation primarily from collision induced hydrogen and helium ionizations and neutralizations. If the proton and helium spatial concentrations were uniform, the pull from gravity would be equal in all directions and the uniform distribution would remain uniform. However, instead this prime-modal universe was rapidly expanding and small in-homogeneities were sufficient enough to cause gravitational clumping. The universe began to acquire a structure by first forming small structures (stars) before larger ones (galaxies). Thus, many of the first generation stars formed first before being grouped into galaxies. Many of these first generation stars are in global clusters some of them still found just outside our Milky Way galaxy in our galaxy halo, having been formed before our galaxy.

Many in these clusters contain red giants in their last evolutionary stage before collapse into white dwarfs or neutron stars. These are groups of first generation stars whose isotopic content is mainly hydrogen and helium and would not be expected to have planets. In our Solar system, all the planets, the four terrestrial and four gaseous (excluding Pluto) have iron cores, not contained in first generation stars.

Chapter 3 – Gravity and a Clumpy Universe on the Small Scale

Cosmologist consider the universe as homogeneous on the large scale since galaxies are distributed reasonably uniform throughout the currently observable regions of our universe. However on the small scale on the order of light years, the mass distributions can only be considered very clumpy with the original matter from the Big Bang being concentrated in individual stars and groups of stars primarily as galaxies. We can attribute this to an extremely dominant force on the cosmological scale called gravity. In between these clumpy objects, the matter density is very low as we will discuss below. Even now, with the overwhelmingly abundant knowledge about high energy physics and our sub-atomic particles, we know very little about gravity. As with many physics related phenomenon, by observations, first by Newton, we understand gravity and have equations to describe and predict it's behavior but do not know it's true makeup. And further we now are aware of another force from recently confirmed Dark Matter mentioned above. Einstein's gravitational theory encompasses fairly simple mathematical equations (the field equations of general relativity) and only becomes complicated when applied to real life astrophysical conditions. It was his life-long ambition to formulate a unified quantum, nuclear and gravitation theory, which even to this day has not been achieved although quantum field theory depicts gravitational force as the exchange of virtual gravitons between two masses just as electromagnetic force depicts the exchange of virtual photons. Even before Einstein's theory of gravitation, Newton showed that two masses. m_1 and m_2, mutually attract each other with a force, F, given by

$$F = m_1 \, m_2 \, G \, /r^2$$

where $G = 6.672 \times 10^{-11}$ Newton meter^2 per kilo-gram^2 is the universal gravitational constant and r (in meters) is the distance between the two masses (in kilo-grams). Further, Newton's second law provided that a mass, acted upon by a force, will accelerate in the vector direction of the force i.e. $F = m\,a$. Thus even the Big Bang protons (neutralized) were accelerated towards each other even though their masses are very small (1.66×10^{-24} g). The force between them at a distance of 1 meter is $F = (1.66 \times 10^{-27}$ kg$)^2 \times (6.672 \times 10^{-11}$ Newton m^2 / kg$^2)/(1$ m$^2) = 1.84 \times 10^{-67}$ Newtons (kg m/s^2). With Newton's 2nd law, we have an initial acceleration of $a = dv/dt = 1.11 \times 10^{-40}$ m / s^2 where v is velocity and t is time. So over a period of 13.7 billion years $= (13.7 \times 10^9 \times 365.25$ d/y x 24 h/d x 3600 s/h$) = 4.32 \times 10^{17}$ seconds, this small amount of attraction results then movement and coalition of the particles. It is most probable however that the coalition was initially driven by random particle collisions. The particle collisions just from the Big Bang effect most likely attributed to the grouping of the hydrogen and helium masses into galaxies and stars within the galaxies.

Chapter 4 – Initial Formation

Initial Massive Stars, Supernovas, Planets, Star Clusters and Galaxy Formations

The first stars then contained primarily only hydrogen and helium, hence were without any chance of having planets since both hydrogen and helium remain a gas down to -259 *C. They were formed from early collapsing gas clouds. The hydrogen proto-stars were formed at a relatively high temperature because no heavy elements were present to cool the gas. Cooling by atomic hydrogen results in relatively massive stars with short lifetimes on the order of millions to hundreds of millions of years. This is borne out by the fact that in our Solar neighborhood there are few stars of low metallicity, most of which would long ago become extinct. This means that in our region, the Solar population are predominantly second or greater generation stars including ours. The massive first generation stars underwent supernova explosive endings. Figure 2 shows the youngest known supernova remnant in our galaxy, G1.9+1.3 recently discovered by Reynolds et al (2008), which is believed to be only about 110 years old. Shown is its remnant debris extent in 1985 (reddish orange region) and its extent in 2007 (greenish region). The supernova is near our galaxy center, has a diameter now of about 6 light years and is expanding at about 14, 000 km/second. Some of the remnants of the iron core are expanding at about 3.8 million km/second.

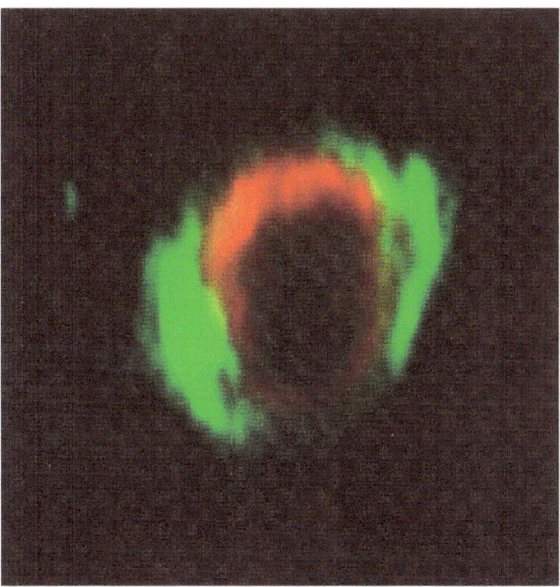

Figure 2, Supernova remnant G1.9+1.3 showing debris expansion extent in 1985 (orange) and 2007 (green)

As we see the speed of ejection from supernovas can be very high. These cosmic particle velocities are energetic enough to extensively produce fusion of heavier isotopes. Figure 3 shows a much older supernova remnant in the Large Magellanic Cloud.

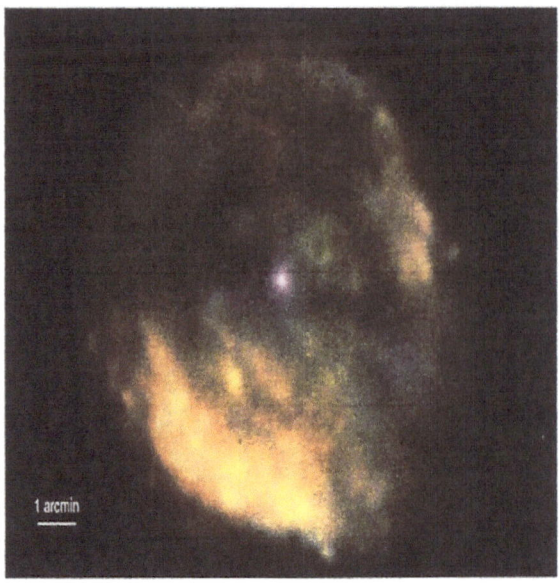

1 arcmin

Figure 3, Older supernova remnant

The spiral shape indicates that the supernova explosion was asymmetric, creating rotational angular momentum to the debris cloud. As Newton showed, angular momentum is conserved as well as energy and linear momentum. Our Solar system initially possessed rotational angular momentum which is still conserved by our Sun and its planets. Just as with the Crab Nebula, this remnant appears to have a residual central star. In some instances with white dwarfs having a companion star, supernovas occur with the white dwarf completely destroyed leaving only the highly energetic, expanding debris. These debris continue to spread creating very large molecular clouds on the order of thousands of pc (1 parsec - 1 pc = 30.9 trillion kilometers or about 3 light years), such as the Crab Nebula now believed to have been created from a 1054 AD supernova.

Stars above about 1.4 Solar masses are capable of undergoing a supernova explosion as its final evolutionary stage. There are two type supernovas, Type I being from the collapse of a white dwarf (Type 1a Thermonuclear explosion) and the Type II being the final gravitational collapse and explosion of a massive star such as was formed in first generation stars. The Crab Nebula, which has a neutron star at its center, is believed to be an example of a Type II supernova explosion occurring as noted in 1054 AD. Supernovas produce nebulae, with their debris, that can occupy vast regions of space up to 2000 light years (LY) in diameter and as little as 1-2 LY in diameter. The nebulae are then the remnants (debris) from the exploding supernova star and are rich in heavy isotopes but abundant in hydrogen and helium also. Thus nebula are breeding grounds for new star formation with potential for planets. Many large nebulae show the formation of and presence of many new stars.

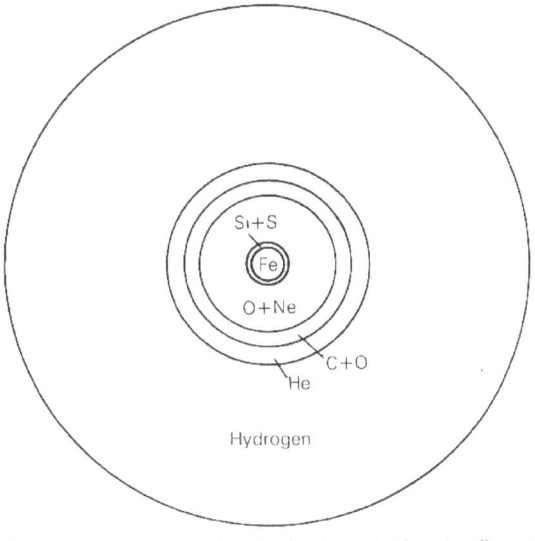

Figure 22.0 Just before its final gravitational collapse

Figure 4, Radial Distribution of Elements Formed in Older Star Near "Burn-out"

Figure 4 shows the "onion" like structure of a massive star before its disintegration showing the distribution of elements as the star reaches the end of its energy generating fusion fuel consumption. Elements greater than iron require input of energy to produce fusion to heavier elements, thus when an iron core is formed and no more energy is generated, the star collapses causing the supernova explosion.

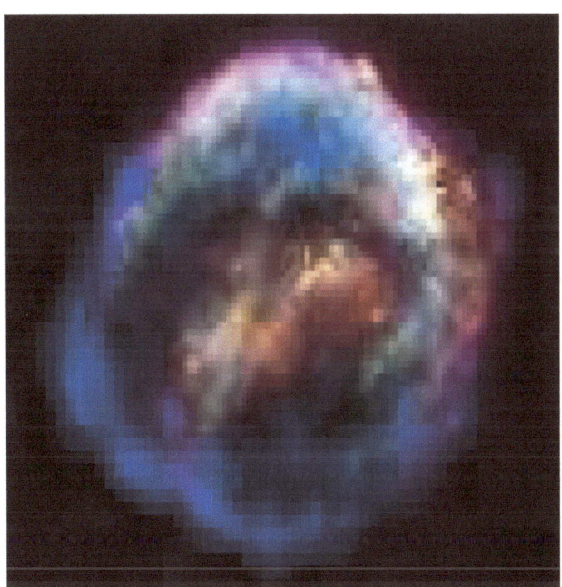

Figure 5, Kepler's supernova seen by him in 1604 AD and as it has expanded today

As Figure 5, we show the spiral shaped Kepler's supernova remnant of 1604 AD located in the Large Magellanic Cloud, with a diameter of about 14,000 light years. The spiral shape is caused by the supernova explosion being asymmetric providing rotational angular momentum to the remnant. The remnant debris expansion over time is modulated by collision with interstellar molecules and the mutual pull of gravity of the debris. The resulting quasi-stable region of debris

11

thus forms a certain type nebulae. The large Crab Nebulae, mentioned above, was formed by a supernova explosion in 1054 AD, reported in Chinese records which also recorded supernovas in 185, 1006 and 1181 AD. Since then Tycho Brahe and Johannes Kepler recorded details of a supernovas explosion in 1572 and as well as the one in 1604 AD. With the Hubble telescope, a number of supernova remnants have been discovered. On January 9, 2008, the first actual modern era observation of a supernova explosion, now known as SN 2008D, was observed including its initial X-ray flash created at the time of the massive stars collapse. SN 2008D is located in the galaxy NGC 2770, approximately 90 million light years from Earth. Figure 6 provides panels showing the X-ray flash event.

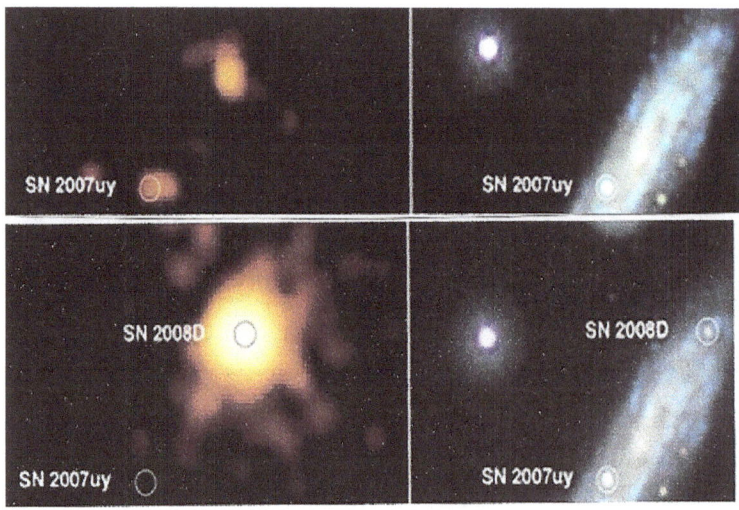

Figure 6, Most recent supernova, SN 2008D, January 2008

The initial star was originally about 30 times the mass of our Sun. The explosion ejected about 7 Solar masses of debris into space and has collapsed into a black hole and a weak, mildly relativistic jet was produced. The upper left panel shows the X-ray view of the star before the supernovae explosion. The upper right shows the visible light image of the region, showing the NGC 2770 galaxy. The lower left panel shows the X-ray flash from the explosion event and as the lower right panel, the visible light spectrum of the region is shown. In these panels is shown the older supernovae SN 2007uy located within another part of the same galaxy. Figure 7 shows the Palomar – 60 inch telescope image taken on 12 January 2008, in visible light. Here the remnant of an older supernovae, SN 1999eh is shown which has faded in intensity, becoming a nebula.

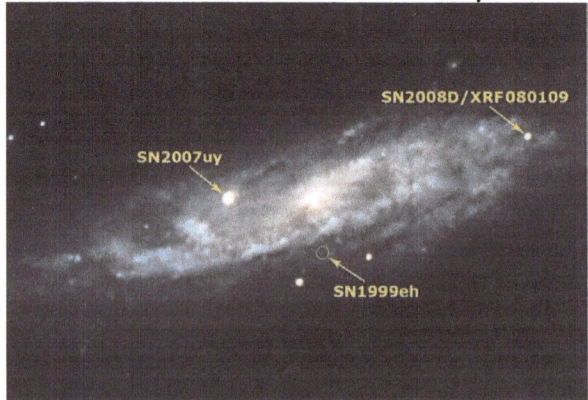

Figure 7, Palomar 60 inch telescope image of SN 2008D in 2008 and surrounding supernova remnants

Chapter 5 – Where are We?

Progeny Stars, Our Solar System in our Galaxy and Planet Formation

As we noted, nebulae are formed from the supernova debris and becomes a breeding region for young stars with heavy element composition and likely candidates for planets. Figure 8 shows the sequence of events after the nebulae becomes quasi-stable.

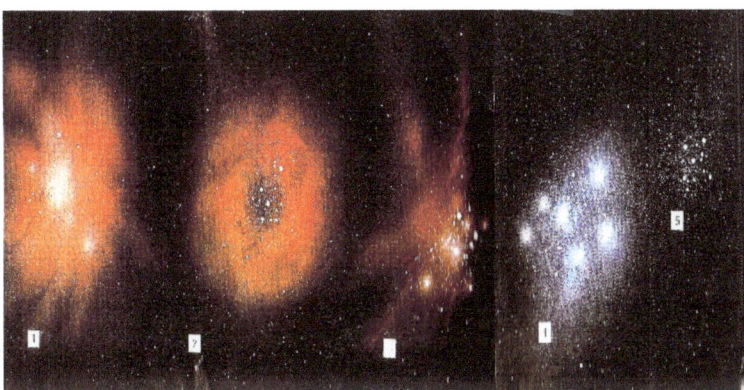

Figure 8, Sequence of events after the occurrence of a supernova showing gravitational formation of star clusters such as occurred in the formation of our second or possibly third generation Solar system

We use the word "quasi-stable" because even the Crab Nebulae is still expanding, but with a relatively low velocity of about 1400 km/s. In panel 1 of Figure 8, we show an initial supernova remnant nebulae after its rapid expansion. Panel 2 shows the nebulae after star formation has begun. Panel 3 shows a nebulae, after most of the expanding debris that is not involved in star formation, has dissipated. Panel 4 shows the region as a star forming region without the presence of the nebulae with supernova debris remaining around the stars, capable of forming planets. Panel 5 shows the debris having being accreted into planets and the remainder being blown into outer space from the Solar winds from each star. Of course the different panels are of different nebula formations and different cosmic regions since the time-scale for the evolution of a supernova remnant through the nebulae stage to the star cluster stage is on the order of millions of years. Remaining in panel 5 then is a cluster of stars with heavy elements capable of forming terrestrial planets with iron cores. We know that our planetary system was formed before our Sun ignited its Solar furnace. We have now identified over 200 stars in our neighborhood with planets. Figure 9 is the image of a proto-star in the Orion nebulae called M42protplyds, about one light year wide, with a proto-planetary disc (nebulae debris cloud), which obviously is in the process of forming one or more planets. The dark center region is the central proto-star that has not yet ignited its thermo-nuclear furnace. We know that our Sun's fusion furnace ignited after the planets were accreted and further its intensity was initially weaker than now.

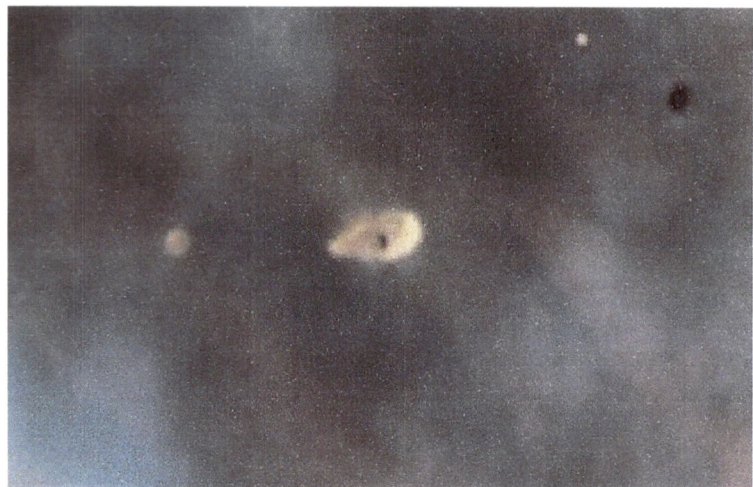

Figure 9, An evolving proto-star, M42protplyds in the Orion nebulae, showing formation of planetary system from nebulae debris

We know that Earth's Sun is at least a second or even greater generation star. We noted that first generation stars are massive, consisting of hydrogen and helium with no chance of planets. The lifetime of the massive first generation stars are known to be on the order of 10s of million years.

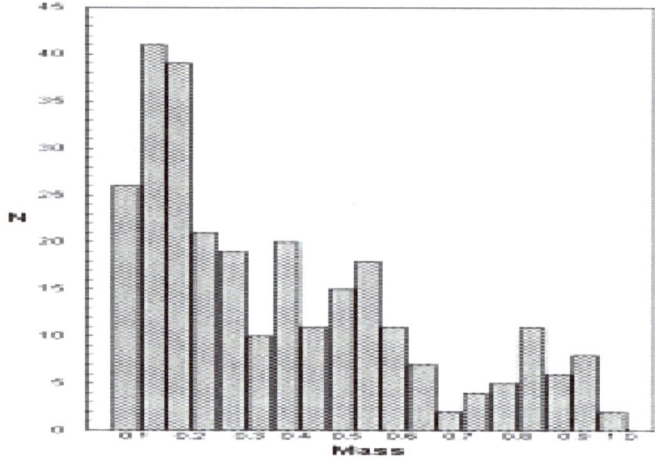

Figure 10, The number mass distribution of stars in vicinity of our Solar system

So our galaxy has had many supernova explosions which has produced vast regions of molecular clouds and nebulae with corresponding turbulence. A study of present stars in our galaxy found that most stars are smaller than our Sun. Edward Salpeter (1955) showed for our present galactic neighborhood that the initial Solar mass (at formation) distribution function, N(M), of present local stars is inversely proportional to a dimensionless constant $\alpha = 2.35$ i.e., $N(M) = M^{-2.35}$, where M is Solar mass. Pavel Kroupa (2012) and others find that solar masses below one Solar mass (Earth mass), is similar i.e. $\alpha = 1.3$ between 0.08-0.5 Solar masses and $\alpha = 0.3$ below 0.08 Solar masses. Figure 10 shows the mass distribution versus stellar mass in Solar mass (Solar mass = our Suns mass) units.

Considering the Oort Cloud as the outer-most gravitational influenced region of our Solar system, then our Solar system is indeed extremely large, about 50,000 AU (AU = distance Earth to

Sun = 93 million miles) or about 0.79 light years. Voyager 1 has just reached the edge of our Solar system, 4 billion miles past the sub-planet Pluto. The closest stars to our Solar system is a triple star system called the Alpha Centauri star system containing Proxima Centauri, at 4.2 LY, and Alpha Centauri A and B, both about 4.3 LY. Alpha Centauri A is slightly larger than our Sun (10%) and Alpha Centauri B is slightly smaller (90% of the Sun's mass). Proxima Centauri is about 12% of the Suns mass. There is no evidence that our Sun is exerting any gravitational effect on these stars. We show the relative sizes as Figure 11.

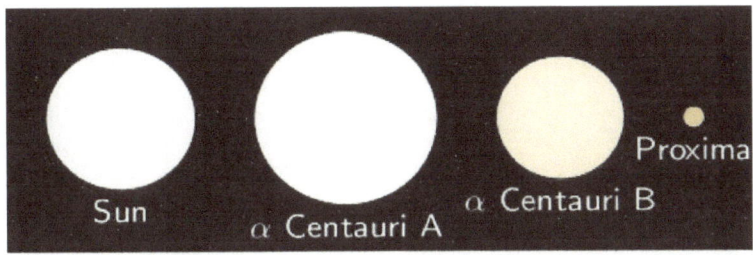

Figure 11, Relative sizes of nearest stars to our Solar system

There are 10 more stars or binary star systems within 10.5 LY of our Solar system including the binary star system, Sirius, between 200-300 million years old and 8.6 LY away. Sirius A is about twice our Solar mass and is the brightest star in Earth's sky due to its close proximity. Sirius B was much larger, about 5 Solar masses, and has thus consumed its fusion energy and collapsed into a white dwarf, with about 1.0 Solar masses collapsed into about the same volume as Earth, about 120 million years ago. Figure 12 shows the distance of the local stars in the neighborhood of our Solar system.

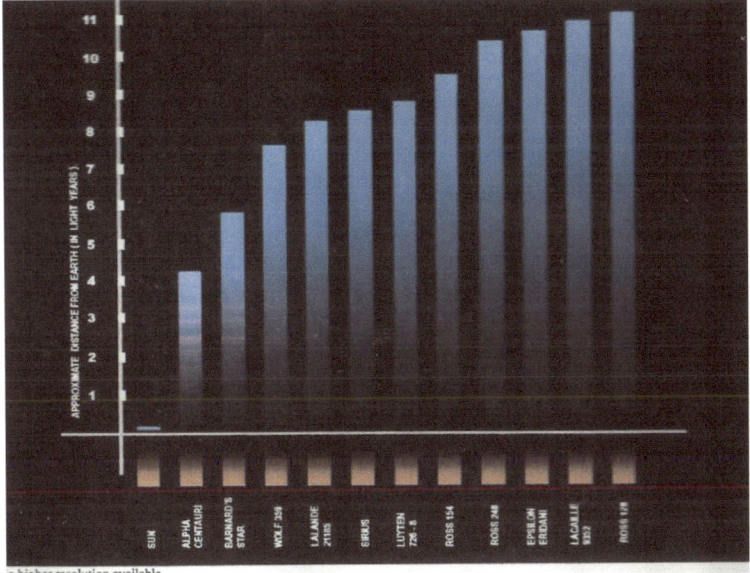

Figure 12, Distances of stars or binary systems in vicinity of our Solar system

There are groups of stars that share a common motion. The Ursa Major Group is a set of 220 stars that move as a common small cluster and with common ages of 500 ±100 million years old. This star group most probably are progeny from a single nebula like shown in Figure 8, panel 5, and thus moving together. Our Solar system does not appear to be a member of a common motion group and is moving through what is called the Local Interstellar Cloud. There is considerable variation in molecular density within our galaxy. The cloud is about 30 LY across and is molecularly very thin, with 0.26 atoms per cubic centimeter; approximately one-fifth the density of galactic interstellar medium but about twice the density of the Local Bubble (a close by low density region).

Figure 13 shows the Local Interstellar Cloud with our Sun and Sirius. This shows that Earth's motion is relatively independent of other stellar objects.

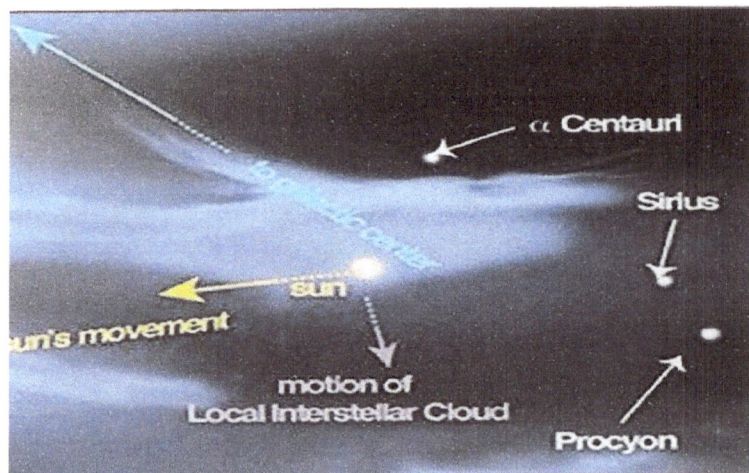

Figure 13, Motion of our Sun with respect to surrounding low density Local Interstellar Cloud

Figure 14 shows on a larger scale the region of the spiral arm, Orion Arm or so called Local Orion Spur, of the Milky Way showing the low density Local Bubble.

Figure 14, Our Sun in vicinity of the low density Local Bubble

As Figure 15 we show the Local Bubble and the Galactic Neighborhood of the Orion Spiral Arm (of our Milky Way galaxy) showing the motion of the Sun.

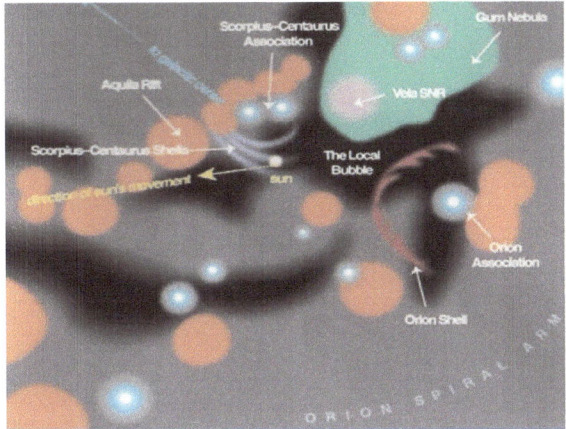

Figure 15, Our Sun in relation to Orion Spur and Spiral Arm

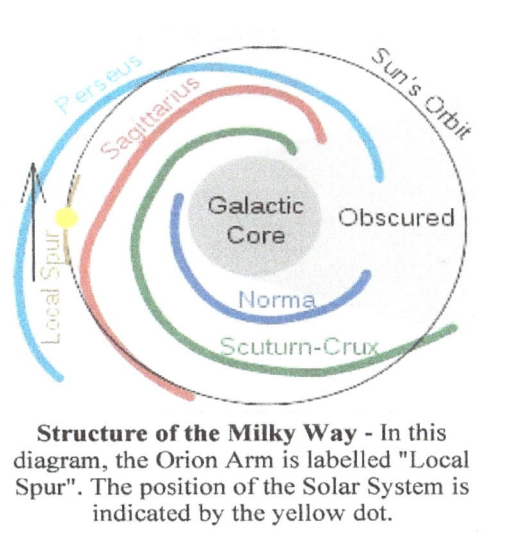

Structure of the Milky Way - In this diagram, the Orion Arm is labelled "Local Spur". The position of the Solar System is indicated by the yellow dot.

Figure 16, Our Sun and the Milky Way galaxy

Figure 16 shows the Orion Arm (Spur) in connection with the neighboring galaxy (our Milky Way) arms and the Galactic Core. Our Sun is moving through our galaxy but also our galaxy is rotating with respect to our universe with a frequency of revolution of 225 to 250 million terrestrial years.

So our Solar system is neither part of a common motion group or a nebula or an interstellar cloud but seems to be independent of these structures. Our Sun is expected to move out of the Local Interstellar Cloud, between 10,000 and 20,000 years from now, into the low density Local Bubble. Ninkovic and Trajkovska (2006) have surveyed the stellar population in the Solar neighborhood and determined their mass distributions. They observed 94 stars within 10 pc and found 50% with mass between 0.1-0.5 Solar masses, 34% between 0.5-1.0 and 16% between 1.0-5.0. Many of the low mass stars are red dwarfs and brown dwarfs. Red dwarfs have mass below 0.4 Solar masses, have low luminosity, and are believed to be the most common star in our galaxy. Planets have recently been found to orbit two nearby red dwarfs, Gliese 581 and OGLE-2005-BLG-3901L. Abad and Vieira (2005) have measured the motion of 5700 stars in the vicinity of our Sun using the Hipparcos parallax method (see Appendix A for the parallax method of cosmic distance determinations) and found that as a group they have a common motion of 25 km/s towards the apex (1,b)=(178.4,4.8). Supernovas occur at the end of the lifetime of massive stars as much as 100 times larger than our Sun. Because they are so massive they burn up their hydrogen fuel much faster and have a much shorter time span. It is estimated that our parent supernova star, before our Sun, lived through its stellar cycles generating heavy metals (astronomers consider any element about helium as metallic.) We know that there are stars in our galaxy that are progeny from third and even fourth generation supernovas. As seen in Figure 4 and 17 below, iron is one of the last fusion products before the supernova event. Fischer and Valenti (2003) used measurements of the iron absorption line to determine the iron to hydrogen ratio in 754 nearby stars. Of these 61 stars had confirmed planets and 693 stars without confirmed planets. The study strongly supports a correlation between star metal abundance and planet formation. Figure 17 provides this correlation between planet occurrence and iron composition in stars.

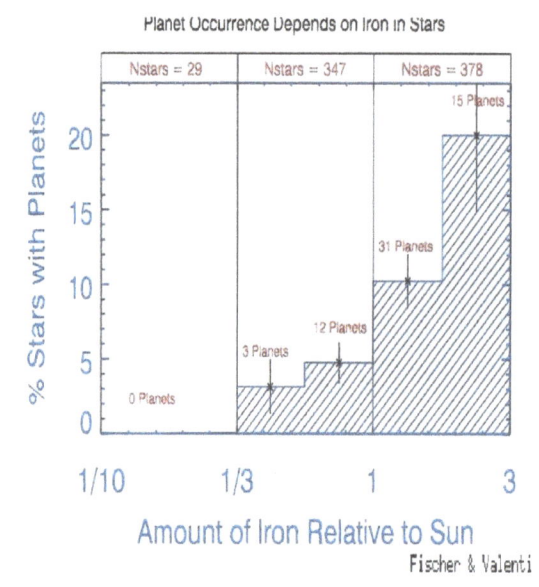

Figure 17, Correlation between frequency of planets and iron composition of parent star core

The histogram shows for the 29 stars with iron content between 1/10th and 1/3rd of our Sun had no stars with planets (0%). At the other end, 20% of the 75 stars with 3 times the relative iron content as the Sun had confirmed planets (15 planets). These stars apparently were formed from debris from higher generation supernovas than our own Sun. We can only speculate about the history before our Solar system but the fact that our system has 8 planets (plus Pluto) may mean that our system is a progeny of multiple generation supernovas as we have mentioned several times before. All of our planets, even the outer "gaseous" ones have iron-rich cores. An unanswered question is why there is no white dwarf or neutron star remnants in our vicinity from our last supernova, although it is known that some supernova explosions are so violent that they leave no stellar remnant. We appear to be alone from the last nebula as suggested by Figure 15.

PART II — THE SOLAR SYSTEM AND FORMATION OF EARTH

In this Part II containing Chapters 6 through 10, we present what we know about the formation of our Solar system, its planets and in particular our Earth. The formation of our Moon will be presented in Part III.

Chapter 6 - The Formation of our Solar system

In our galaxy we see billions of examples of star formation as occurred with our Sun so we know a lot about what happened in our own Solar systems evolution. Figure 18 shows the distribution of the elements in our Solar system including the Sun, the planets and other asteroids, comets and interplanetary dust.

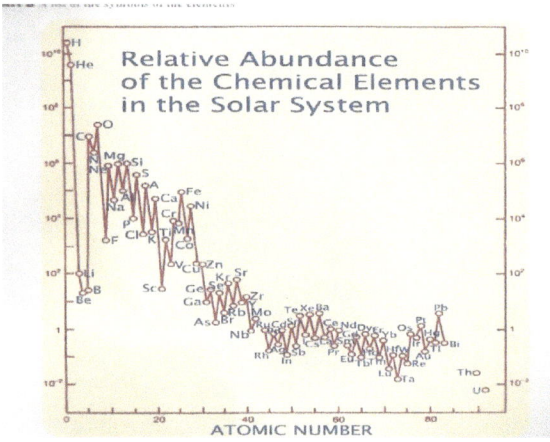

Figure 18, Distribution of Elements in our Solar System

It is seen that the exogenous fusion products Li, C, O, Si, Ca, Na and Fe are orders of magnitude more abundant than the heavier elements, in monotonically decreasing amounts relative to their atomic numbers. What goes on inside the stars relative to generating fusion energy is complex (Harms et al 2000). The internet Wikipedia, the free encyclopedia, provides an excellent presentation. Basically, it is the ultimate in example of Einstein's equation, Energy = Mass c^2, where c is the speed of light. In each fusion process, two (or more) atoms combine to form a new, heavier atom, but weighing less than the original two atoms. This lost mass is converted into energy i.e. Gained Energy = Lost Mass x c^2. In simple terms, this is how stars and our Sun acquire their enormous energy. In the case of the fusion of two protons, since they each carry a positive charge, their Coulomb repulsive must be overcome, this on the Sun by a tremendous internal pressure (this repulsion is why we have not yet harnessed fusion energy here on Earth). But this is why stars and our Sun's fusion furnace is slow being ignited. The first step in the stars fusion process is the fusion of two protons into a deuteron and a positron. The higher elements above Fe are much less abundant because they are only primarily created from the brief high energy collision (endogenous fusion) period during the expansion of the debris from the supernova explosion. Most of the original matter prior to the explosion of our massive parent star is blown away into outer space with velocities of thousands of km/s. Initially only the debris with angular momentum and other debris obtaining tangential velocities gained from mutual particle collisions during the expansion would not be lost in space (the radial ones would be quickly lost at extremely high velocities). If the explosion was asymmetric then the entire residual matter left behind would have some orbital angular momentum, as was the case for our Solar system. This is why the Sun rotates and all the planets orbit in the same direction – counter-clockwise looking down from above the North Pole.

The idea of our Solar system forming in relative isolation from the debris of a single supernova was preferred up to a few decades ago. Recent studies (Knie et al 2004) of ancient meteorites have revealed traces of short-lived iron-60 which only forms in exploding stars. This

20

suggests that perhaps a number of supernovas exploded before and while the Sun was still forming in a giant molecular cloud. A shock wave from perhaps the first supernova may have caused the compression of the local cloud region and creating a region of over-density and causing a pre-Solar small nebulae (several light years across) which then gravitationally collapsed into our Sun. This occurring 5 billion years ago, would suggest that our Sun may have been formed in a large star-forming region where massive, supernovae type stars are produced. The closest such region is the Orion Nebula (24 light years across). The Orion molecular cloud, partly shown in Figure 15 containing the Orion Nebula, is several hundred light years across. The most dense region of the Orion Nebula, called the Trapezium, has about 2,000 young stars in an area of about 4 light years diameter (about the distance to our Suns closest neighbor). Figure 9 provides a photograph of M42proplyds, a proto-planetary disk around a star in the Orion Nebula. If created in Orion Nebula, our Sun has since moved away some 1200 light years (which would be at a nominal cosmic velocity of about 0.26 km/s), perhaps initially propelled by the supernovae shock wave.

From the observation of star formation in our galaxy, it is known that within 50 million years after the beginning of the collapse of our solar nebula, the temperature and pressure at the core of the Sun became great enough that the hydrogen began to fuse into deuterium and helium, creating an internal source of energy which countered the intense gravitational force, prevented our Sun's further contraction and enabled achievement of stellar hydrostatic equilibrium. This marked the entry into the Sun's prime phase of its life, the main sequence phase, which is the phase that stars derive their energy from fusion of hydrogen into helium in their cores. Our Sun will remain in the main sequence for about 5 billion more years.

The planets were formed from gas and dust in the solar nebula cloud surrounding the Sun as it was forming – Figure 9 shows such a cloud for M42proplyds. The formation, called accretion, began as dust grains in orbit around the central protostar. Through direct collisions these grains formed clumps between 1 and 10 meters in diameter which in turn collided to form larger bodies (planetesimals) of 5 km. These were large enough to exert gravitational pull on the grains and clumps and other planetesimals and also with the random collisions grew at a rate of perhaps centimeters per year for the next few million years. As noted, the planets acquired their near present size before the Suns fusion furnace began. After between three and ten million years, the young Sun's Solar wind cleared away all the remaining gas and dust in the protoplanetary disk. However, even some of the solar nebula gas constituencies, mainly hydrogen, were accreted on the so called gas planets Jupiter, Saturn, Uranus and Neptune. All four of these giant planets have iron cores but otherwise are almost completely metallic and liquid hydrogen due to the large pressure of gravity and colder surface temperatures. Figure 19 shows the delineation that occurred (and most likely occurring in Figure 9) as our Solar system was being formed where beyond the "Frost line" the so-called gas planetesimals retained iron cores and liquid and solid (ice) hydrogen and some helium. The inner terrestrial planetesimals initially had hydrogen and helium as a gas envelope. In fact all the planets initially had hydrogen and helium atmospheres with iron cores. This will be discussed in a later section.

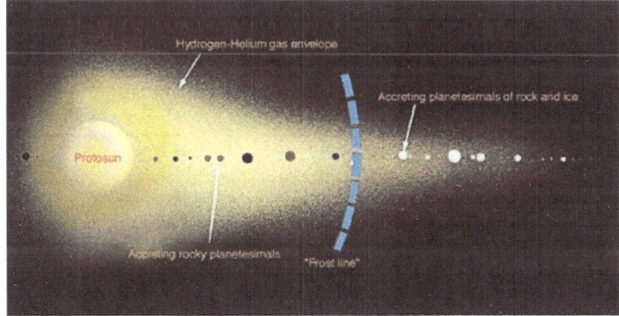

Figure 19, Formation of terrestrial and gaseous planets

Chapter 7 - Formation of Planets, Earth and Moon

Accretion of the Solar System and Observed Properties

As noted above, the general consensus is that our Sun is a second or third generation medium sized star which was formed out of the debris from a supernova explosion of a first or second generation star. If a first generation star, it was probably formed as long as 12 to 13 billion years ago soon after the big bang. Thus it would have contained mostly hydrogen and some helium (from the big bang and fusion during its lifetime) and would not have contained an abundance of the higher massed elements that make up the inner terrestrial planets of our Solar system. Our predecessor star would most likely have been void of any planets. The supernova explosion of our predecessor produced numerous combinations of nuclear fusion type reactions to produce our current abundance of higher massed elements that we find today from Lithium to Uranium plus some even higher massed elements and their isotopes with shorter half-lives (compared to the age of our Solar system $\approx$ 5 billion years). Figure 18 shows the abundance of isotopes in our Solar system. The most well-known naturally radioactive isotopes are Uranium isotopes 235 and 238 and Potassium 40 still present due to their relatively long half-lives of 7.1×10^8, 4.51×10^9 and 1.26×10^9 years, respectively. The Radon and its progeny, from the Uranium decay chains and the ^{40}K are the major source of humans background radiation exposure. There are about 40 other naturally occurring radio-isotopes still present with measured half-lives ranging from 10^{10} to 10^{20} years.

We do not know whether our Solar system is the only secondary star formed from the remnants of our precursor first generation super-nova star. Outside our planetary system but still a part of our Solar system, we have two regions of other planetary objects. The Kuiper Belt and the Oort Cloud in a region from 50,000 to 100,000 AU distances from the Sun (about 1 light year). The estimate of the total mass of the cloud is rather small, between 5 and 100 Earth masses. This outer Solar system region is the source of comets that are in orbits but with some, from time to time, knocked to cross into Earth's path by mutual comet collisions. As noted above, our closest neighbor is Alpha Centauri, 4.2 light years away with no evidence of a planetary system. This suggests that our Solar system is the only progeny of the first generation star and any other material has dissipated into intergalactic space. As noted above, Figure 15 suggests that our Solar system is moving through our galaxy alone. According to the Chandrasekhar limit (the maximum mass of a stable white dwarf star due to inner electron collapse), only stars of mass greater than 1.4 times our Sun can develop into the supernova explosion condition. We know of stars with 15 times our Sun. So at least 40% of the original first generation stars has been lost. The escape velocity from our present Sun is about 600 km / s. We can observe a number of Nebula left over from supernovae and in modern times have experienced several. Scientist at the Las Alamos National Laboratory Supernova Science Center estimate the initial particle radial velocities to exceed 4,000 km/s, so it is not unreasonable to assume that considerable mass from the first generation star was lost to intergalactic space. Any explosion would be expected to produce an inhomogeneous cloud of debris, not only in mass distribution but also with motion such that a net rotation (angular momentum) would be expected from asymmetric supernova explosions. Due to Newton's conservation laws, the cosmic cloud formed after the explosion would also conserve any rotational angular momentum of the original first generation star also. These individual embryonic particles that remained in the solar system space would have radial and tangential motions with respect to the gravitation center of the Nebula cloud, where the Sun is forming, and will be in Kepler orbits around the condensing Sun [two of Kepler's laws of planetary motion state that 1.) the orbits of the planets are ellipses, with the Sun at one focus of the ellipse. 2.) The line joining the planet to the Sun sweeps out equal areas in equal times as the planet travels around the Sun.]. As the debris accreted into planets, this rotation was conserved, thus all planets orbit around the Sun in the same direction, as expected and noted above. The Sun itself rotates in this same counter-clockwise direction (looking down on our North Pole) with a period of about 26 days. Although there are many premises regarding the details relative to the evolution of our Solar system, this heterogeneous theory is generally accepted.

Chapter 8 - Early Accretion of the Nebula Particles

Recent work has described the planetary accretion of the inner Solar System (Chambers 2004). After the super-nova explosion and the remnants achieved stable Newtonian movement, the Nebula consisted of gas molecules and dust which migrated into a protoplanetary disk (like the rings of Saturn) around the Sun due to the sun's vertical component of gravity and conservation of angular momentum. These typical dust grains were estimated to reach this midplane disk in about 10^4 years. At first the collisions of these small sized particles caused accretion growth by only the particles sticking together or otherwise fragmenting depending of the relative velocity of the collisions. Particles with collision speeds of up to tens of meters per second would stick together, others would be fragmented. This non-gravitational accretion eventually resulted in the formation of planetesimal of a few km in diameter where gravity then provided some accretion influence. This resulted in most collisions leading to net accretion, unless the impact speed is substantially higher than the targets gravitational escape velocity or the impact is a grazing one. The larger planetesimals feed on the smaller ones and the larger ones become planetary embryos. These with their substantial gravitational pull quickly outgrew all others resulting in what is called the runaway growth phase of our Solar accretion. Each of these embryos staked out a feeding zone of our Solar disk. The final stage of accretion involved only a few dozen embryos with masses on the order of 0.01 to 0.1 Earth masses (Moon to Mars sized). This is illustrated in Figure 19. Chambers estimates that the period from planetesimal to completion of the runaway growth to be from 0.1 to 1 Million years. The total time for final formation of Earth and the other planets is estimated to be as little as 5 Million years but as much as 100 million years. More than the present eight planets (excluding Pluto) are premised to have been accreted i.e. source of our Moon impact object and large moons on the gaseous planets and an impact object on Mars and possibly Venus and Uranius. With this large relative uncertainty, it is difficult to estimate when and where the "Giant Impact" Object (GIO) for our Moon formation (Hartmann and Davis 1975, Cameron and Ward 1976) became influenced by Earth. Our calculations below beginning in Chapter 11 provide estimates.

Chapter 9 - Time Sequence - Formation of Solar System, Formation of Planets

The age of the Sun, now determined to good accuracy is about 4567.4 ± 0.5 M years. We above presented a rather dismal picture of estimating the time sequence for the formation of the Solar System. Recent isotopic analysis of meteorites, chondrites and CAIs (calcium aluminum rich inclusions) have now provided reasonably accurate data. The oldest solids (other than cosmic dust) in our Solar System are the CAIs. Fortunately, three isotopes, 41Ca, 26Al and 53Mn are useful with half-lives of 0.1, 0.7 and 3.7 M years, respectfully. We can readily assume that these isotopes were formed by fusion processes at the time of the super-nova explosion i.e. the beginning of our embryonic Solar System Nebula. The radio-isotope dating reveals that the CAIs were formed within several hundred thousand years of the beginning. Similarly, we find that the larger chondrites, with CAIs embedded in them, were formed about 4564.7 ± 0.6 M years. This gives an interval of 2.7 ± 1.1 M years between the formation of the CV CAIs and the CR chondrules (Amelin et al 2002). More recent analysis (Amelin 2005) of a separate type of chondrules (CB), with two age values of 4562.7 ± 0.5 and 4562.8 ± 0.9 M years, that could have only been formed by highly energetic impacts with large objects at least the size of planetary embryos. This provides an indication that the Solar System had passed the planetesimal stage and reached the planetary embryo stage by about 5 M years. Chambers notes that the asteroids took only about 2 M years to form. We can therefore estimate that the planets were formed as early as 5 to 10 M years after the birth of the Solar Nebula, this including Earth and the "Giant Impact" Object. The asteroids and meteorites were formed billions of years ago, they have been estimated from the ages obtained from radioactive dating of the oldest meteorites. The very oldest of the meteorites would have been formed in the pre-planetesimal period, which could be as early as 1 Million years after the formation of the Nebula cloud. It is estimated that hydrogen burning in the Sun began about 40 million years after the Nebula was formed.

The Initial Accretion "Feeding Zones" for Earth and the "Giant Impact" Object

There are various theories relative to the source of the "Giant Impact Object" (GIO) relative to the formation of the Moon. We have chosen to locate the GIO as initially accreted between Earth and Mars at an outward distance of 40 M km during their initial accretion stage. If we assume the mass of GIO to be the same as Mars, then the boundary where the gravitational forces are equal would be 30.14 M km outward from Earth and 9.86 M km inward from GIO. The mass density of the gravitational "feeding" region around GIO before accretion must include the region outward from GIO towards Mars.

Chapter 10 - Why Don't All Planets Rotate in the Same Direction

A simple explanation has not been offered as to why our planets rotation axis (axial inclination with respect to its orbit around the sun) varies so greatly (Giuli 1968, Ohtsuki 2012). A predominance of the planets rotate counterclockwise (West to East). Only two, Mercury and Jupiter have their axis of inclination near zero, Mercury 0° and Jupiter 3.12°. Mercury's rotation is retrograde (East to West). Coincidentally or not Earth, Mars, Saturn, and Neptune are all about 24° to 30° West to East (Earth 23.5°, Mars 24°, Saturn 26.75°, and Neptune 29.56°). Venus is very non-conforming with an axis inclination of 178° and so is Uranus with an axis of inclination of 97.8°. One explanation of how the planets developed their axial rotation is to compare their movement through the nebula debris during their accretion period. When the planets were first forming it would be with the faster moving debris making the most accretion collisions by catching up with the slower moving, making more collisions and sweeping up these slower moving ones. If the debris density was slightly greater towards the Sun then the planet would develop a counterclockwise rotation. To examine this hypothesis, we have estimated the radial density variation outward from the sun. For each planet we computed the region of space where their gravitational pull equaled that of their next neighbor, which would be the accretion "feeding space". This would provide an estimate of the radial dependent mass distribution of the Solar system. For Mercury it would be, towards the Sun, the radial distance where its pull equaled the Sun's and, away from the Sun, the radial distance where its pull equaled that of Venus. With these regions of gravitational dominance, we have computed the volumes of the regions and, with the mass of each planet, the estimated debris density in the region while the planet was forming.

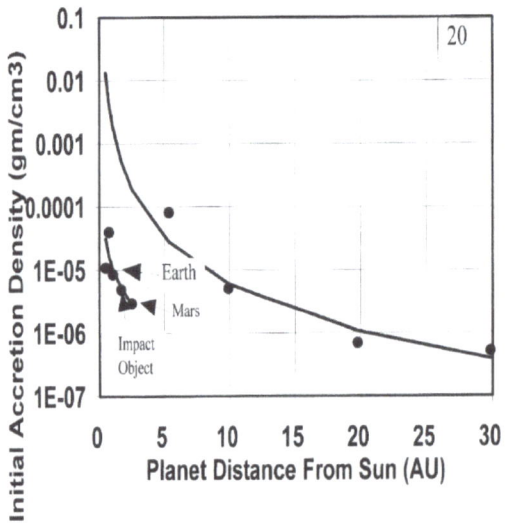

Figure 20, Based on Orbital Locations of Venus, Earth and Mars and Premising a Monotoncally Decreasing Cosmic Density, the Initial Location of the Sister Planet Impact Object after Accretion is Shown

As Figure 20, we show the estimated mean density within the gravitational dominance regions about the planet versus radius from the Sun in AU units. We show also good agreement with the radial density distributions predicted by the accretion density model of Bogojevic et al (2006). We see that between Mercury and Venus we have a density increase. This is most likely because, close into the Sun but even past Venus, the Solar wind (blow off) was so intense when the Sun "turned on" its furnace that the microscopic initial debris was depleted. This radial negative density gradient around Venus would mean a greater collision rate on the outer side of the planet away from the Sun and thus an unequal force tending to rotate the planet clockwise. The density is monotonically decreasing with radial distance for the other terrestrial planets and for the outer giant gaseous planets past the Asteroid belt tending to result in counterclockwise rotation, except for Venus and Uranus which both may have experienced Giant Impacts, as the accreted particles struck the planets in an uneven distribution inward towards the Sun. We note that the axial rotational "day" on Mercury is 58.6 Earth days, which suggest minimal density gradient during its formation with perhaps the Solar wind being equal on its "back" and "front" sides. Venus's Solar "day" is even longer, 246 Earth days, where we find from Figure 20 a lesser density gradient than Mercury, so its retrograde rotation is insignificant. Mars Solar day is 24.6 hours suggesting the density gradient to be significant and to have been similar to ours during its accretion period, inducing nearly the same axial rotation period.

The Axial Inclinations

We consider the rotational periods for Mercury and Venus to be negligible, their axial inclinations are insignificant and so is the 3.12° of Jupiter. One would suspect that axial inclinations of Earth, Mars, Saturn, and Neptune, all between 23° and 30° must be correlated with some common feature of their accreted formation. Improbable hypothesis would premise that collisions with comparable sized, massive bodies, or a sudden bombardment with an avalanche of large asteroids from an explosion in the asteroid belts could produce this "tilt" with respect to their Solar orbits. In the latter case all four planets at nearly the same time.

PART III — A QUANDARY RELATIVE TO FORMATION OF THE MOON

In Chapters 11 through 14, we begin with the present consensus of geophysicists that the Moon was formed by the collision of Earth with a Giant Impact Object (GIO). It is not at this point in time certain as to the size of the GIO, its angular momentum, the impact velocity and the angle of impact. With all these variables to contend with it is not odd that there have been many theories about the formation of the Moon. Key contributors to these GIO theories are Dr. A. Cameron and Dr. R. Canup and their associates. The size of the GIO has been thought to be on the order of the size of Mars but recently a study suggests the GIO to be of equal size to the pre-impact Earth (Canup 2012). Also unresolved has been the origin of the GIO prior to impact (in other words "from whence did it come"). We have concluded that only if GIO was created by accretion into close orbit with Earth would there be a reasonable probability that an impact would occur. The premised size of the GIO also strongly suggests its accretion as a "sister planet" close to Earth. We present the calculations relative to this impact. We show that a near-Earth accretion, rather than identical orbit accretion proposed by Gott (2011) is most probable.

Chapter 11 – Theories on the Origin of the Moon

Solar astrophysicists are now in agreement on one major aspect of the formation of our moon, the so-called "Giant Impact" scenario i.e. that the Moon was created by a colossal collision with a Mars-sized proto-planet object, which presumably occurred towards the end of or after the planetary accretion process. Figure 21, Panels B through E illustrate the "Giant Impact" event.

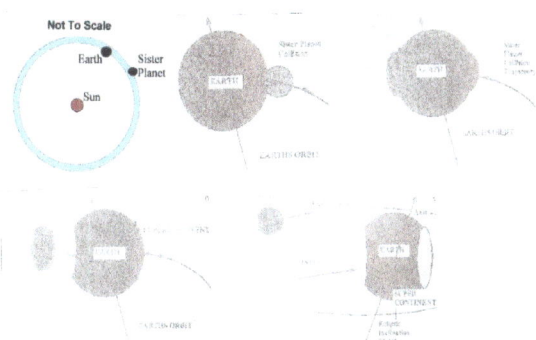

Figure 21, Geophysicist Consensus that Moon Was Created by a Collision of Earth with a Giant Interstellar Object

The Giant Impact hypothesis was first premised by Hartmann and Davis (1975). At about the same time Cameron and Ward (1976) had the same idea. Now the consensus of the Solar astrophysics community is that the Giant Impact model is the only plausible explanation for the Moon being what it is today (Canup and Esposito 1996, Ida et al 1997, Cameron and Benz 1991). Compelling evidence shows the Apollo Moon samples have the identical oxygen isotope ratios as Earth, to a very high procession, indicates that Moon material was accreted at the same orbital location as Earth or that the Moon material was originally Earth material. The most recent evidence to support the near-Earth orbit Giant Impact Theory is of Paniello et al (2012) showing zinc isotope vaporization at time of impact. Zinc is strongly fractionated when volatized in planetary rocks, but not during normal volcanic igneous processes. Moon rocks contain more heavy isotopes of zinc, and overall less zinc. This is consistent with zinc being depleted from Moon through evaporation, as would be the case for the Giant Impact after collision.

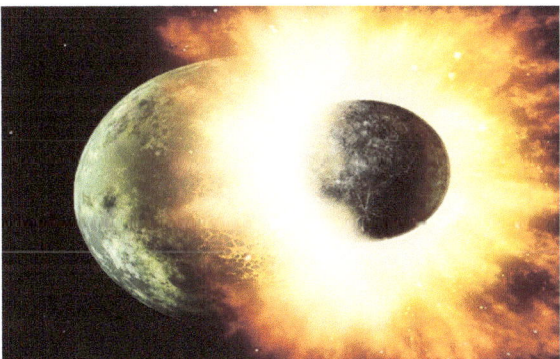

Figure 22, NASA/JPL-Caltech artist's Impression of Planet Sized Collision to Form Moon

There are three factors associated with the formation of the Moon as a satellite of Earth. One is that the Moon contains a significantly less iron in its core than Earth as if formed mainly from mantle and crust material. The second is that its oxygen isotopic concentration directly matches Earth's again indicating it was formed nearly in the same orbit as Earth or of Earth material as suggested below. Third, the rotational angular momentums, as well as energy and linear momentum, before and after the collision must be conserved as given by Newtonian mechanics. We show as Figure 22 an impression of the collision of the possible Giant Impact Object. The impact collision has been suggested to be a glancing one (but, this would mean more iron in the moon's core) with a large part of the Mars-like proto-planet shearing off and the fragments forming a vaporized ring beyond the Roche Limit and accreting into our present Moon but initially much closer, perhaps only 75,000 miles away. (The Roche Limit is the distance within which a celestial body, held together only by its gravity, will disintegrate due to a second body's tidal forces exceeding its own gravity.) A number of investigators have examined the kinetics of the collision and post-collision Moon formation including a four part presentation by Cameron and his colleagues (Benz et al 1986, Benz et al 1987, Benz et al 1989, Cameron and Benz 1991). Recent computations by Canup (2012) have found that the collision of two nearly equal objects, in his calculations, more closely satisfies issues such as angular momentum. Three conservations must be satisfied i.e. Energy, linear momentum and angular (rotational) momentum. For any Giant Object collision, most of the material being ejected would be vaporized from the heat generated from the kinetic energy of the massive object striking at as much as 11.2 km/second (the escape velocity from Earth). In Figure 21D, some of the vaporized material will escape from Earth's gravity field. There are so many variables in postulating the detailed construction of an "origin of the Moon" scenario i.e. size of GIO, impact velocity, GIO angular momentum, size of proto-Earth, angular momentum of proto-Earth. As noted a surprise was the evidence from the Apollo Moon samples that Moon and Earth have the same oxygen, tungsten, chromium, and titanium isotopic signatures, strongly showing that the Moon and Earth were at least accreted in the same very close neighborhood.

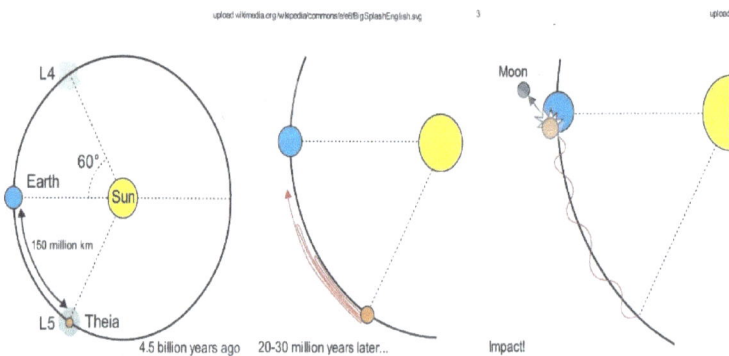

Figure 23, Gott (2011) theory that the Moon accreted at same orbit location as Earth

In addressing the isotope signature issue, Gott (2011) has proposed a theory that the Moon was accreted at exactly the same position in the Solar system as Earth. He proposes that the Moon accreted at the same orbit as Earth at a stable Lagrange point either 60 degrees ahead or 60 degrees behind Earth or a separation of 150 million kilometers as shown in Figure 23. We reason that in that case the Moon and Earth would be locked into the same orbital velocity and the gravitational pull at

30

that separation would be minimal, meaning a very long time before the collision occurs. Further, they would have the same accretion feeding zones, hence their formation separately at the same orbit would be unlikely.

After Gott's (2011) premise of a GIO accretion in the same orbit as Earth to remedy the isotopic signature problem, Cuk and Stewart (2012) have presented a model proposing a fast-spinning proto-Earth with a period of 2.3 hours. They propose that the GIO impacted (velocity 20 km s^{-1}) at a head-on or slightly retrograde angle. This results in the GIO material being totally absorbed into the Earth and the ejected Moon material being primarily Earth crust with only some iron core. This is compatible with our model given below.

Due to the uncertainties noted, the pre-collision phase of this Giant Impact theory however has not been analyzed in detail. We proceed here using what we consider reasonable probabilities and examine the possible origin of this Mars-sized "projectile". Because the kinetics of Newtonian mechanics precludes a closed form solution of the multi-bodied problem, we use a perturbation approximation to model the dynamics of the time span prior to the collision. We further, examine characteristics of the Earth and Moon, their properties relative to our Solar system and premises for the origin of these characteristics and other properties relative to the formation of our Moon.

Chapter 12 – The Giant Impact: Moon Formation

The Giant Impact Relative to Earth's Moon Formation

Our Moon is the largest planetary satellite, relative to the size of its planet, in the Solar system (excluding the Charon-Pluto system). Others have estimated that the size of the Giant Impact Object, that is premised to have resulted in the formation of the Moon, would be on the order of Mars. A fundamental question is, "Where did this massive object come from?" We must assume the Giant Impact object (GIO) originated within our Solar system. The GIO is larger than any asteroid by over a factor of 80. It is larger than any planetary satellite other than those of the giants Jupiter (4 greater) and Saturn (4 greater). It is more massive than Pluto by a factor of four. The GIO is therefore so large that the matter comprising it must have been accreted as planetesimal material and in a conventional planetary orbit configuration, during the Solar systems embryonic accretion period.

Then the question is whether the accretion took place in some other region of the Solar system and by another predecessor giant collision, been deflected towards the Earth just as we know that asteroids , meteorites and comets are deflected in our direction by collisions with other objects that prior were in the conventional stable Solar orbits. The probability of the GIO collision with Earth in one pass, for a GIO knocked to cross Earth's path, is about one in 20,000. For two crossings about one in 10,000. So it is unlikely that the GIO was a giant asteroid in the asteroid belt redirected by collision with another giant asteroid. First for this to occur the prior collision would have to be between our GIO and a second planetary sized object. If this happened, where is this other second object now? This would suggest the very remote probability of two giant collisions, the first to deflect the GIO towards us and second for the GIO to collide with Earth as it crosses Earth's orbit. This occurring most likely as a single crossing or at most, infrequent crossings. Canup (2004) showed that in the final stages of the Solar system formation, about 20 planet embryos were formed from the smaller planetesimals, such as illustrated in Figure 19, and it was at this stage that the Giant Impact Object forming our Moon occurred. By mutual giant impact collisions, the final four terrestrial planets were created with one being from the Giant Impact collision to form our Moon, another to perhaps explain Mar's thin atmosphere and another to explain Venus's retrograde rotation and another to explain °Uranus' 90° axis. The only scenario with any reasonable probability then is that the GIO accreted in a conventional solar orbit in the vicinity of Earth's orbit. This is similar, but slightly different from Gott's (2011) theory, in this way this GIO, which we could call a companion "sister" planet, would have many close encounters, passes, with Earth because they would have different orbital velocities. The head-on premise of Cuk and Stewart (2012), we will show below is compatible with our scenario.

Chapter 13 – Initial Solar Orbit

Most Probable Initial Solar Orbit for an Accreted Impact Object

We have determined that the Giant Impact Object was most probability formed in a Solar orbit in the vicinity of Earth's orbit. We show this as Panel A of Figure 21. The mass of the GIO and Earth's mass must be at least equal to the current mass of the Earth and the Moon. This would be a lower limit estimate since after the massive collision with Earth some vaporized mass would be expected to be lost to other Solar regions or drawn into the Sun. We, for our initial computations consider the total pre-collision mass to equal that of the present Earth and Moon.

We shall now endeavor to compute the scenario resulting in the eventual collision between the GIO and Earth. An important unknown is the initial orbital configurations of the GIO "sister" planet and Earth. But before we address this unknown, we will perform some preliminary orbital computations. We shall consider a geometry with the Earth as the origin of a moving frame of reference around the Sun. The pre-collision orbit of GIO we assume to be approximately in the Earth's orbital plane and it's orbital inclination nearly zero. We feel this to be justified since Mars's orbital inclination is 1° 51' and Venus's is 3° 24'. For simplicity we also assume that both orbits are circular, like shown in panel A of Figure 21. We first examine the kinetics of the last stages of the two objects orbit behavior prior to the ultimate collision. The sidereal period of a planet is a function of distance from the Sun. The sidereal period can be derived as follows

$$\text{Gravitational Force} = Fg = G\, Ms\, Mp\, /\, r^2 \qquad (1)$$

where G is the universal gravitational constant $= 6.67 \times 10^{-20}$ km^3 / s^2 kg, Ms and Mp are masses of the Sun and the rotating planet and r is the radius of the planets orbit. The centripetal force of a rotating body is given by

$$\text{Centripetal Force} = Mp\, v^2\, /\, r \qquad (2)$$

where v is the orbital velocity. For a stable orbit to be maintained these forces must be equal, i.e.

$$G\, Ms\, Mp\, /\, r^2 = Mp\, v^2\, /\, r \qquad (3)$$

Now v is related to orbital (sidereal) period, T, by $v = 2\pi r / T$
Solving for T we obtain

$$T = 2\pi r^{1.5} / (G\, Ms)^{0.5} = 0.172 / r^{1.5} \qquad (4)$$

if r is in M km units and T is in days. We see that T is independent of the mass of the planet.

We wish to next compute the time between passes for the Earth - GIO orbital system. We will assume that the GIO's orbit is outside the Earth's orbit, towards Mars. As a first approximation, we assume the pre-collision orbit of Earth is the same as now with a radius, r = 149.5 million kilometers (M km). With to GIO further away from the Sun than Earth, Earth will have a shorter sidereal period (365.25 days) and will be continually catching up and passing the slower GIO. With each pass, there will be a gravitational encounter where the mutual attractive force will gradually pull the two masses closer to each other, the smaller GIO being pulled inward to a slightly smaller orbit with each passage. We can compute the time between passes versus orbital separation distance. The difference between the orbit periods is given by

$$\text{Delta T} = \text{GIO Period} - \text{Earth Period} = 0.2 \left[\left(1 / Re \right)^{1.5} - \left(1 / RIO \right)^{1.5} \right] \qquad (5)$$

where Re = Earth radius and RIO = GIO's radius.

The number of revolutions for Earth to catch up and pass GIO is given by

$$\text{Time per pass} = \text{GIO Period} / \text{Delta T} \qquad (6)$$

Chapter 14 – Gravitational Deflection of the Impact Object Orbit per Pass

In the Appendix B, we formulated a perturbation model to compute the gravitational deflection of the Sister Planet (SP) Giant Impact Object per pass between SP and Earth from the mutual gravitational pull during the time that the two objects were within each other's gravitational field. If we assume that Earth's displacement is negligible relative to the displacement of the much smaller SP due to its much smaller mass (1/10th of Earth), we can compute the gradual approach of SP towards Earth with each passage. We have examined a number of initial orbitals for SP. We consider the most probable position of SP after the run-away accretion period to be 40 M km outside Earth's orbit, or a total distance from the Sun of 190 M km (This is an assumption to facilitate our calculations.). The total circumference of this orbit would be C = 2 π x 190 M km = 1,190 M km. In the model it is necessary to decide at what point in the Earth's orbit to begin the displacement calculations (the value of Xo in the calculations). We have decided to use Xo = - 30 M km to begin with Zo = 30 M km = distance SP orbit outside Earth's orbit. When the planets are pulled together such that Zo = 10 M km, we change Xo to 20 M km.

We next must estimate the orbital spacing of these two planets, obviously between the orbits of Venus and Mars. We have above computed the present gravitational regions (the regions where each planet is gravitationally dominant) for the planets in estimating the regional mass densities during the accretion period. For present day Earth and Moon, we found this to be 20.4 million kilometers inner region boundary and 73.1 million kilometers outer boundary. The laws of Newtonian mechanics provides that the radial distance of a planet from the Sun is independent of the mass of the planet, just dependent on the orbital velocity i.e.

Centripetal force = Suns gravitational force

$$m v^2 / R = G Ms\ m / R^2 \qquad (7)$$

such that

$$v = (G Ms / R)^{1/2} \qquad (8)$$

The orbital velocity is also given by v = 2 π R / P where P is the period of Solar revolution (for Earth P = 365.25 days).

Again G is the universal gravitational constant = 6.67×10^{-20} km^3 kg^{-1}s^{-2} and the mass of the Sun = 2.0×10^{30} kg. Then G Ms = 1.33×10^{11} km^3 s^{-1}. This would satisfy our requirements also.

So the orbital velocity depends on the orbital radius, R.

A Probable Initial Location for the Impact Object

The upper limit for the mass of the Impact Object is estimated by most geo-scientists to be comparable to the mass of Mars which is about 1/10th the mass of Earth. In Figure 20 we showed the estimated mass densities of the Nebula cloud before significant accretion began. With the mass of the Impact Object being much less than Earth, we would premise that it must have accreted in a region of much lower mass density, thus outside the Earth's orbit. The GIO also however must be close enough to Earth that Earths' gravitational pull dominates over that from Mars. Both planet's orbits are slightly elliptic but the average distance between the orbits is about 78 M km. To obtain Figure 20, we computed the boundaries between the planets where the gravitation forces are equal. The Earth-Mars boundary is about 52 M km from Earth and about 26 M km inward from Mars. We rather arbitrarily chose the GIO position as 30 M km outward from Earth and a distance of 48 M km inward from Mars. At this location, the gravitational force from Earth is 19.9 times that of Mars on GIO. In Figure 20, we show the location of GIO with respect to Earth and Mars.

The Gravitational Force, Relative Orbital Velocities and Time Between Passes

With Earth inward towards the Sun from GIO, Earths period of revolution around the Sun will be shorter than GIO's such that Earth will be passing GIO, moving ahead and catching up again from behind GIO on subsequent revolutions. The initial relative orbital velocities, Vr, is given by

$$Vr = Ve - Vio = (G\ Ms)^{\frac{1}{2}} [\ (1\ /\ Re\ )^{\frac{1}{2}} - (1\ /\ Rio\ )^{\frac{1}{2}}\ ]$$
$$= (G\ Ms\ )^{\frac{1}{2}} [\ 0.000045 - 0.0000279\ )$$
$$= (1.33\ \times 10^{11}\ )^{\frac{1}{2}} [\ 0.000017.1\ ]$$
$$= [\ 3.64 \times 10^5\ ] [\ 0.0067 - 0.00528\ ]$$
$$= 5170\ km\ /\ s$$

Figure 24 provides a graphical presentation of the relationship between the Sister Planet and Earth.

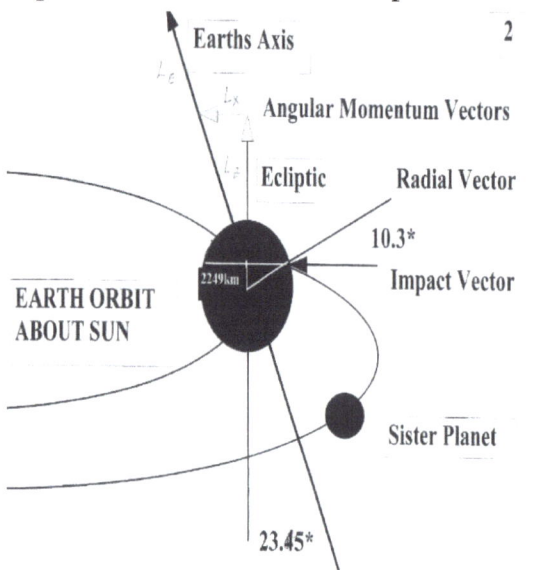

Figure 24, Graphical Illustration of Relationship between Earth and Sister Planet

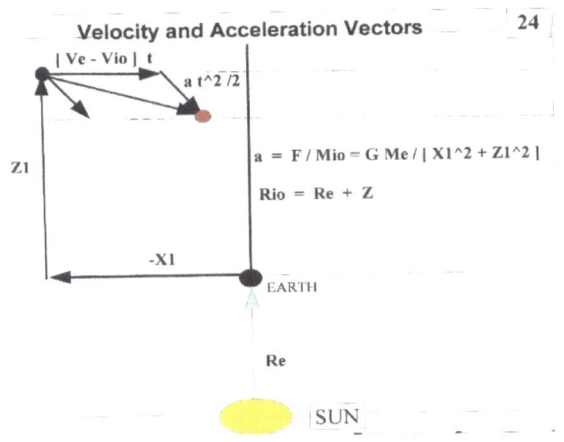

Figure 25, Model Velocity and Acceleration Vectors for Sister Planet Passage of Earth

To further show the models parameters and the Earth and Sister Plant relationship, Figure 25 shows the velocity and acceleration vectors for the calculations.

In the model, for a given increment of time, Δt, the Sister Planet experiences a gravitational force from Earth and is accelerated towards Earth and moves an incremental distance towards Earth. This results in an incremental increase in gravitational force for the next Δt. Figures 26 through 30 show the resulting changes from successive passes for the calculations discussed above. For the Sister Planet to be finally drawn into an impact with Earth required 515,000 passes. Figure 26 provides the accumulated time for the Sister Planet to experience the impact was 18.48 million years.

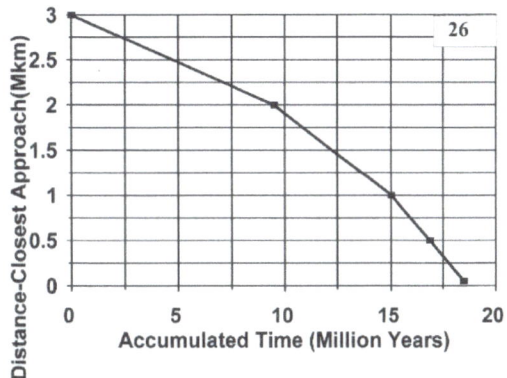

Figure 26, The Accumulated Time for the Sister Planet to Experience Earth Impact

In Figure 27, we show the time required per passage of Sister Planet by Earth as a function of relative distance between the Sister Planet and Earth. This graph shows that the closer the two are to the same object the less the relative velocities and the longer it takes to catch up and pass.

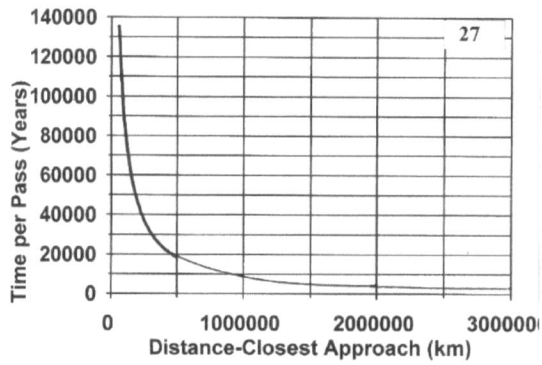

Figure 27, Time Between Passes, Earth and Sister Planet

Figure 28 provides the distance between the Sister Planet and Earth as passes are accumulated beginning with the SP being at a distance of 3,000,000 km apart. It is seen that at that distance the approach is very slow with many passes required. The other graphs show however that the time per pass increases drastically as the two objects get closer since their orbital velocities become nearly equal.

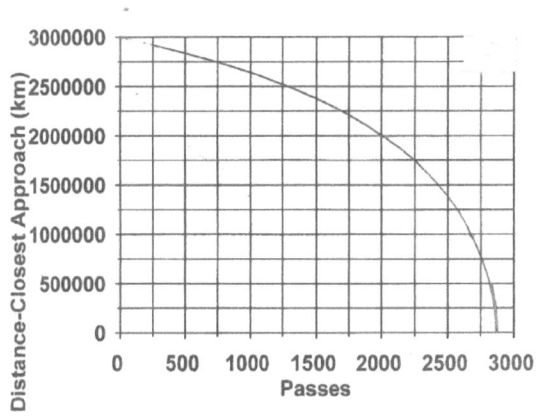

Figure 28, Distance of Sister Planet from Earth as it Approaches as a Function of Passes

38

To illustrate how the Sister Planet monotonically approached the Earth with each successive passage, Figure 29 shows the progress of the approach with the few last passages. This shows that the closer the planets get orbit-wise the greater the amount of change per pass due to greater gravitational attraction.

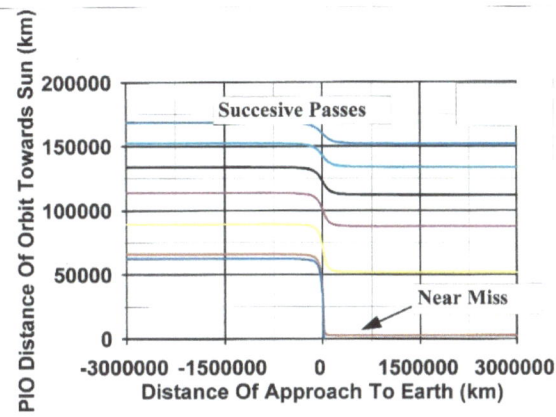

Figure 29, Successive Passes as the Sister Planet is Drawn Towards Earth

Finally as Figure 30, the last two passages are shown i.e. the next to the last near miss and the last impact passage.

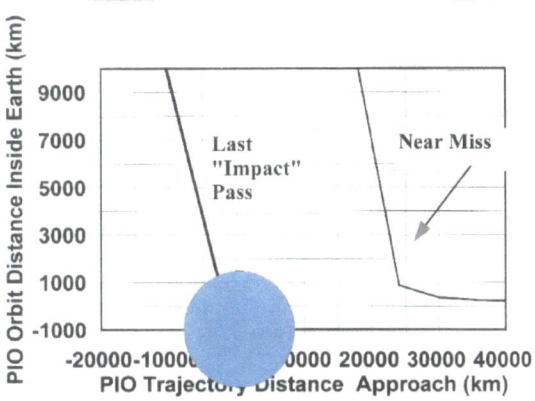

Figure 30, The Sister Planet Path for Last Two Orbits Past the Earth. Final Orbit Produced the Giant Impact Collision Depicted in Figures 21 and 22

Most Recent Apollo Moon Rock Data – Earth Isotope Signatures

Until recently it was assumed that the Giant Impact Object originated some distance from Earth's orbit. Further, others have made the assumption that the Moon's composition consisted of primarily the Impactor's terrestrial material and minimal Earth crust and mantel material by virtue of a glancing collision. However, as mentioned above, Wiechert et al (2001) analyzed Apollo lunar samples and found that the isotope abundances of the oxygen isotopes were identical to Earth's oxygen fractionation. This implied that Earth and the Giant Impactor were formed at about the same heliocentric or the Moon is made up of mainly Earth material. Researchers at Caltech

39

(Pahlevan and Stevenson 2011) argue that it is extremely unlikely that (probably <1 percent) the Impactor and the Earth would have identical experienced isotopic signatures. In an earlier paper they argue (Pahlevan and Stevenson 2005) that after the Giant Impact, Earth and proto-moon experienced diffusive equilibrium with respect to oxygen isotopes and their signatures would be the same. Providing what is considered as conclusive evidence of same signatures, Koppes (2012) reports the work of a University of Chicago graduate student, J. Zhang, that the isotopes of titanium in the Lunar rocks identically match the titanium isotope distribution on Earth. They premise then that the progeny Moon has no other parent, just Earth. For this premise to be true, they suggest the origin of the Moon would have to be by the fission process (a rapidly spinning Earth expelling a large mass of its own material), which is improbable for several reasons. What these isotopic results do however mean is that the Impact object was accreted in an orbit close to Earth's as depicted here but also the Moon being mainly of Earth material ejected during impact. In our Figure 21 above we show the Giant Impact expelling a mass of material into space. Assuming both the Sister Planet and Earth having iron cores, if the collision was a direct head-on (0 degree) impact (i.e. the angle of impact nearing zero with respect to the radial vector of both objects, GIO and Earth), then the consequence would basically be an impact between the two iron cores since the yield strength of their crust and mantel materials would be essentially negligible compared to the iron-nickel cores. As an extreme simplification, we refer to a device that has been called the Newton Cradle, pictured as Figure 31. It consists of two or more identically sized steel balls suspended on a metal frame, perfectly aligned and in contact with each other.

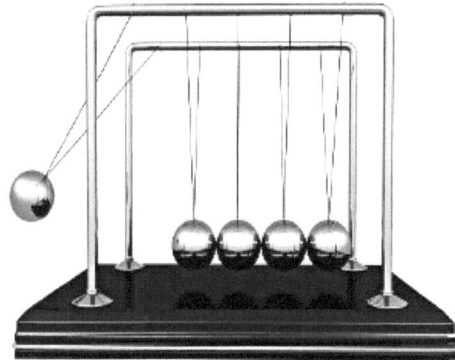

Figure 31, Photograph of a commercially available Newton's Cradle device

In Figure 31, we show five identically matched balls in the supporting frame. If one outer ball is pulled away and is let to fall, it strikes the first stationary ball in the series and comes to a dead stop. The end ball on the opposite side acquires most of the velocity (and energy) and almost instantly swings in an arc almost as high as the release height of the first ball. The intermediate balls remain stationary. The final ball receives most of the energy and momentum that the striking ball lost. Only a small amount of energy is lost to heat within the balls in each cycle. The first ball and last end ball have their energy briefly stored as potential energy when at their maximum height. This commercially available device sells as a toy but also as a physic demonstration educational tool. It is rightfully termed a Newton device because it demonstrates his laws on motion that energy and momentum are conserved. The two end balls would continue to swing back-and-forth if there were no energy loss from heat and friction. Each time the balls are struck the kinetic energy is stored as potential energy by compression of the steel in the balls and this is passed from one to the next. Steel has a high modulus of elasticity and thus little energy is lost to heat (however, some lost each collision).

In the case of the pre-moon giant impact object, if the collision is a direct head-on zero degree one, the collision is primarily iron composition core to iron composition core. Much of the initial kinetic energy will liquefy and vaporize the collision material of both parties to the collision. With the head-on collision, all of the mass of the much smaller Mar-sized Giant Impact object will be absorbed into the Earth's mass. However, the other of Newton' laws, momentum, must be conserved i.e. mass x velocity. Momentum is a vector quantity, in this case the initial momentum i.e. vector quantity $Mass_{Impact\ Object}$ x $Velocity_{Impact\ Object}$. Thus, to satisfy momentum conservation after the collision for the giant impact object just as for the fifth steel ball, material must be ejected from the far side of Earth as rather crudely shown in Figure 21. This means the sum of the ejected particles and vapors masses times their vector velocities (in the direction of the pre-collision GIO) must equal the pre-collision vector momentum. With the Impact Object being mostly fully absorbed internally into the proto-Earth, the ejected material will be nearly all proto-Earth material (from the other side of the Earth) and will have the same oxygen signature as post-collision Earth because it is Earth material. Further, the ejected material, that would constitute to coalesced Moon material, would be low in iron core material since it would be largely Earth crust and mantle material that is ejected. As we have reviewed the various attempts by others to describe the collision and afterward the formation of the Moon itself, before the isotope signature issue was revealed many scenarios were developed. Our computation includes the origin of the GIO as having been accreted in the vicinity of Earth and the pre-collision mechanics of the GIO gradually being drawn towards the Earth's orbit with each Earth passage, resulting in a near head-on collision and the Moon material being nearly all Earth material solving the isotope problem. In the Cuk and Stewart case their impact parameter b = sin ろ = -0.30, this providing impact at 17 ° off Earth's radial vector or near head-on. One significant consideration in all the other studies is the satisfying the angular momentum conservation. As mentioned above, Cuk and Stewart (2012) proposed a fast-spinning Earth could produce a moon-forming disk of Earth's mantle material, where Earth would subsequently lose angular momentum by orbital resonance between the Sun and Moon. This would satisfy our requirements also.

The next question to address is "What is the likelihood of a head-on collision or at least a near head-on collision?", which was also premised by Cuk and Stewart (2012). In Figure 24 we show a vector diagram of relative velocities i.e. the orbital velocities around the Sun and the accelerated velocities towards each other from mutual gravitational pull of the Sister Planet and Earth. In Figure 30 we show that the orbit of the GIO on the next to last passage being about 62,000 km outside Earth's orbit. We estimate that the relative orbital velocity to be 0.549 km/second or about 2000 km per hour (the relative velocity for the last pass with impact will depend on the difference in orbital distances from the Sun. Just happened in pass by pass calculation to be about 62,000 km) Then the main component of the impact velocity would be from the mutual gravitational pull, which would mean a very near head-on collision since the pull is center of gravity to center of gravity i.e. GIO center to Earth center or small angle from their radial vectors.

PART IV — EARTH AND ITS ATMOSPHERES

In this Part IV covering Chapters 15 through 17, we continue the examination of the evolution of Earth. First we describe the formation of Earth's atmosphere beginning with the initial atmosphere of hydrogen and helium and the culmination with our present mixture of primarily nitrogen, oxygen and argon. We show that because Earth unlike Venus and Mars has liquid water, the carbon dioxide that Venus has is scrubbed out on Earth by precipitation. This leaves what is actually a serious deficit of CO_2 for flora but leaving nitrogen like the other two planets since nitrogen is relatively chemically inert and having oxygen from the evolution of abundant flora.

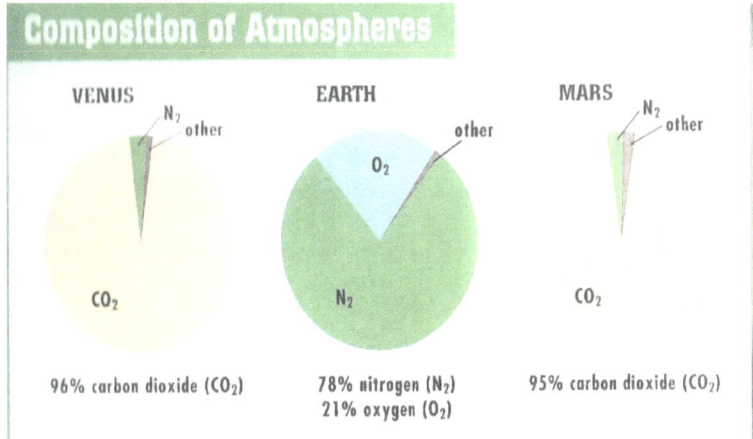

Figure 32, Present Composition of Venus, Earth and Mars Atmospheres

Earth's atmosphere underwent three stages of development beginning with the initial accretion formation, along with the rest of our Solar system, about 4.6 billion years ago.

To examine these stages, we can begin by comparing in Figure 32, the present diverse atmospheric compositions of our Earth and our two neighboring planets, Venus and Mars which in our Solar System have the greatest terrestrial similarity to our Earth. One would first observe that the present chemical compositions of the atmospheres of these three planets are quite different, as if there was never any correlation between the three planets atmospheres. Venus has a very dense atmosphere, 92 times that of Earth, consisting mainly of carbon dioxide (96.5%); Mars has a very thin atmosphere, 0.7% as dense as that of Earth but, like Venus, consisting also mainly of CO_2 (95.32%). Another extreme difference is that neither Venus nor Mars has any obvious evidence of surface water whereas Earth has a hydrosphere of about 1.4×10^{21} kilograms of water.

Oddly to imagine at first glance however, it is believed that initially, during accretion, all three planets formed an atmosphere of hydrogen and helium, which were, and still are considering our Sun, the most abundant elements in our Solar system, and indeed the universe. All three planets were initially accreted from the solid elements in Figure 18 as planetary core and crust and accumulation of the gaseous hydrogen and helium as its atmospheres from the nebula created by a supernova explosion. Being of low mass, these light elements have greater kinetic energy of motion per unit atmospheric temperature and were able to randomly achieve the escape velocities of 10.36, 11.18 and 5.03 km/s, respectively for Venus, Earth and Mars. This is unlike the large planet Jupiter, with its escape velocity of 60.22 km/s, that has kept an upper atmosphere of 82% hydrogen, 17% helium and 1% other heavier compounds such as ammonia, thus being able to retain its initial hydrogen and helium to this day. On all three terrestrial planets then their atmospheres became void of these two lightest elements. The planets were initially molten from the heat of accretion impacts and many of the heavier gasses within the molten material readily escaped from the slowly solidifying crust into the slowly voiding atmospheres from escape of hydrogen and helium. Later as the planets cooled and formed solid crusts, these heavier gasses, such as in Table 1, continued escaping from the crust into the atmospheres, even to this day on Earth, by volcanic and fumarole activity.

Volcano Tectonic Style Temperature	Kilauea Summit Hot Spot 1170°C	Erta' Ale Divergent Plate 1130°C	Momotombo Convergent Plate 820°C
H_2O	37.1	77.2	97.1
CO_2	48.9	11.3	1.44
SO_2	11.8	8.34	0.50
H_2	0.49	1.39	0.70
CO	1.51	0.44	0.01
H_2S	0.04	0.68	0.23
HCl	0.08	0.42	2.89
HF	---	---	0.26

Table 1, Volcanic Gases

Table 1 provides examples of volcanic gas compositions in volume percent's for three studied Earth volcanoes (Symonds et al 1994) showing the major constituents as CO_2 and H_2O (steam).

The second atmospheric stage for all three planets was acquired by the accumulation of primarily CO_2 and H_2O from the volcanos as shown in Table 1. So this second stage of each began with all three planets having nearly the same atmospheres of CO_2 and water vapor. On Venus, the greater Solar intensity maintained the H_2O but only as water vapor without ever a liquid hydrosphere. With the intense Solar photon bombardment causing radiolytic disassociation of the water into H and O, the hydrogen was again lost to space. The heavier free oxygen and high temperatures produced an oxidative environment with oxygen reactions with terrestrial corrosive metals such as iron and also free carbon, forming more CO_2. Thus, Venus has become locked eternally into its present atmosphere containing predominately CO_2, unlike Earth, with no hope of a future liquid water hydrosphere to "scrub out" the CO_2 by Calcium and Magnesium reactions presented in equations (9) and (10) below.

There have been several theories as to how Mars lost its second stage atmospheric and surface constituents of predominantly CO_2 and H_2O. The surface water would be primarily in the form of ice at the average temperature between -20 and -60°C. A currently most accepted explanation is that Mars atmosphere and surface water were blasted away into space when struck by a very large asteroid (Melosh and Vickery 1989) or another Sister Planet as premised above for Earth's GIO. They suggest an asteroid as small as 3 kilometers in diameter is capable, on impact, of exploding into a plume of hot gases that can expand faster than the planet's escape velocity thus sweeping up the surrounding gas in the atmosphere.

We have discussed above the most prevalent premise that the Earth was struck in a similar way by a very large object to form the Moon. The Mars impact theory is supported by the fact that in a large region of Mars surface, the crust thickness decreases from about 50 km to about 20 km over a large area. This hemisphere sized crater is the most visible feature on Mars (Minkell 2008). Mars is known to have had water, at least briefly, since the small amount of hydrogen has excessive isotopic deuterium, the heavier isotope of hydrogen, from the preferential escape of the lighter [1]H

44

isotope. The second atmosphere on Earth, beginning about 4.4 billion years ago when the crust cooled, was still heavily populated with volcanoes which released steam, carbon dioxide, sulfur dioxide (SO_2) and ammonia (NH_4). Like the present Venus, this second Earth atmosphere is believed to could have had an atmospheric pressure approximately 100 times present level. The greenhouse effect from the large amount of CO_2 kept the Earths hydrosphere from freezing solid early in the Solar systems history while the Suns fusion furnace was gaining intensity. A fundamental question then is "What happened to this large amount of atmospheric carbon dioxide here on Earth?" The answer is quite obvious. Falkowski et al (2001) has estimated the present distribution of carbon compounds on the Earth in Giga-tons (1 Giga-ton = 1,000 million tons). This includes all forms of carbon on Earth i.e. all the carbon in the atmosphere, the ocean, the lithosphere, the terrestrial biosphere, the aquatic biosphere and the fossil fuels.

Pools	Quantity (Gt)
Atmosphere	720
Oceans	38,400
Total inorganic	37,400
Surface layer	670
Deep layer	36,730
Total organic	1,000
Lithosphere	
Sedimentary carbonates	>60,000,000
Kerogens	15,000,000
Terrestrial biosphere (total)	2,000
Living biomass	600-1,000
Dead biomass	1,200
Aquatic biosphere	1-2
Fossil fuels	4,130
Coal	3,510
Oil	230
Gas	140
Other (peat)	250

Source: P. Falkowski, et al. 2001: The Global Carbon Cycle: A Test of Our Knowledge Of Earth as a System. *Science* 290, 291-296.

Table 2, World distribution of Carbon

Table 2 provides their data. The major component is in the lithosphere sedimentary carbonates, which are primarily $CaCO_3$ (calcite) and $MgCO_3$ (dolomite). In many places on Earth these carbonates measure thousands of meters thick. In Italy there is an entire mountain range made of dolomite. We have computed the mass of atmospheric CO_2 if all the Earth's carbon was initially atmospheric carbon dioxide. Based on the Falkowski et al numbers, we estimate that there would be 1.996×10^{20} kg of CO_2 in our atmosphere. If we added this CO_2 to our present Earth atmosphere, 98.0% would be carbon dioxide, comparable to Venus and Mars. This value is obtained using the present Earth atmosphere of 5.136×10^{18} kg but subtracting the current mass of atmospheric oxygen — 1.07×10^{18} kg – which was not present due to initial early consumption of any free oxygen by oxidation of Earth's surface minerals i.e. iron and sulfur. As noted, this is comparable to the CO_2 percentages currently on Venus and Mars and what would be expected in the second early Earth atmosphere. Earth most likely did not accumulate all of this CO_2 ejected during Earth's early volcanic episodes but as CO_2 accumulated it most likely was also simultaneously washed out by precipitation. We however show data below that Earth did acquire an atmosphere with CO_2 levels 100 to 1000 times our current low levels (Pagani et al 2005), even after life was thriving on Earth.

We currently have about 2.0×10^{14} kg of CO_2 in our atmosphere. This is extremely small compared to the past, even "recently", i.e., the Carboniferous, Triassic, Jurassic and Cretaceous

45

Periods spanning from 360 to 65 million years BP. In these periods prehistoric flora and fauna life flourished on Earth, as shown below in Figure 33 (Berner and Kothavala 2001) with high CO_2 levels. The most variable components of the Earth's CO_2 in Falkowski's table are the relatively small atmospheric pool and the ocean's vast inorganic pools. Carbon dioxide solubility in water varies inversely with water temperature so increased ocean temperature from a global warming climate will increase the atmospheric CO_2 component but with a temporal lag (such as coming out of the recent ice ages). This will be extensively examined and discussed in later sections dealing with the Global Warming issue.

Several investigators have modeled the time variation of atmospheric carbon dioxide considering the various mechanisms which influence each of the carbon reservoirs listed in Table 2. Three such models are BRYOCARB (Fletcher et al 2005), GEOCARB (Berner and Kothavala 2001) and the analytical model of Rial (2004). Figure 33 provides the most recent version of the GEOCARB model showing estimated atmospheric CO_2 levels from 600 million years ago to present.

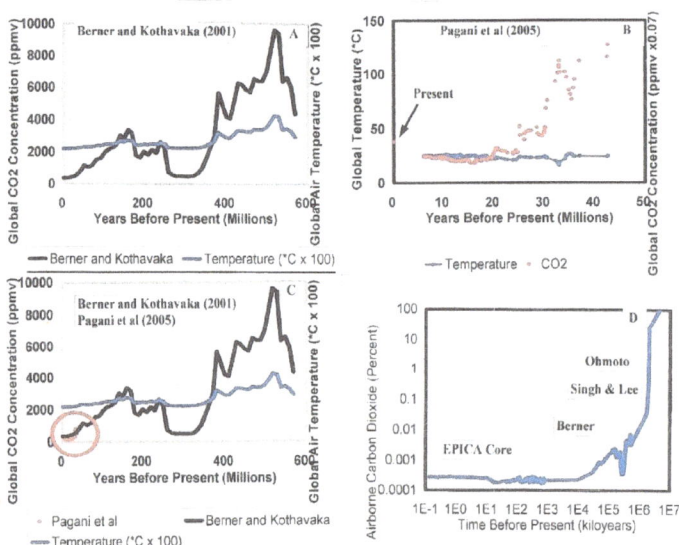

Figure 33, Past variations of Airborne Global Carbon Dioxide Concentrations based on the GEOCARB modeling calculations of Berner and Kothavaka (2001)

(Panel A, with measured data of Pagani et al (2005), Panels B and C, and Ohmoto et al (2004), and Singh and Lee (2007), Panel D with the assumption that the initial Earth pool of carbon given in Table 2 was airborne CO_2.)

The Figure 33A data of Berner and Kothavala are in good agreement with the data of Ekart et al (1999) with levels as high as nearly 10000 ppmv at about 950 Ma BP based on radioactive dating of carbonate soils. Figure 33B, shows the atmospheric CO_2 data of Pagani et al (2005) for the period from about 45 million years before present, which are in relatively good agreement with Berner and Kothavala. As Figure 33C, we provide the Berner and Kothavata and Pagani et al data together showing a good agreement in the recent times as CO_2 receded from the extremely high values and approached its present, extremely low, quasi-equilibrium level between 190 ppmv and 380 ppmv. Little data are available on CO_2 levels in very early Earth before 600 Ma BP. Ohmoto and Kazumazawa (2004) provide CO_2 levels at 1.8 Ga BP (1 billion years = 1Ga and BP - before present). Singh and Lee (2007) provide levels at 36 million years BP and Lowe and Tice (2004) estimate CO_2 levels at 3.2-3.0 Ga to be 100-1000 times present. Figure 33D shows the estimated

CO_2 levels based on these very early estimates.

An alternate theory of early Earth has been recently proposed by Kopp et al (2005). The hypothesis proposed is that on early Earth, with the Solar luminosity at about 80% of its present value, the atmosphere contained both CO_2 and methane, with the strong greenhouse warming provided by the more effective greenhouse gas, methane. The early bacteria are believed to have evolved to be dependent on soluble iron or hydrogen sulfides (and not CO_2). This form of synthesis similar to that of organisms recently observed in the deep-ocean vents at the mid-ocean Atlantic ridge. The origin of CO_2 consuming cyanobacteria, which came later, using photosynthesis to convert CO_2 to O_2, not only reduced the greenhouse atmospheric CO_2 but also produced atmospheric O_2 to react with the methane, thereby drastically reducing its global warming effect and trigging the first ice age. This is the "Snowball" Earth estimated by Kopp et al to have occurred about 2.3 billion years ago producing Earth temperatures as low as -50 °C. Evidence of the "Snowball" glaciation at low latitudes is found in sediments of the Makganyene formation of South Africa which was in tropical latitudes at that time due to migration of the early super-continents. Without some change in the Solar irradiance, the Earth could have been locked into an eternal, highly reflective (high albedo) "Snowball". The Kopp et al premise is that Snowball Earth was thawed by a resumed greenhouse gas effect from volcanic production of CO_2 over a period of millions of years. Most geo-scientist favor the high CO_2 and methane early greenhouse theory for early Earth.

Chapter 16 – Atmospheric Oxygen Evolution & Presence of Nitrogen, Argon

Oxygen being highly reactive, for billions of years did not appear as free molecular oxygen in the atmosphere. The volcanic gases consisted primarily of CO_2 and H_2O, as shown in Table 1. Some radiolytic disassociation of water vapor by Solar UV light would produce free O_2 which would immediately react with iron and other corrosive metals exposed on the Earth's crust. In this period, about 3.4 billion years ago, first life is believed to have appeared. First life consisted of anaerobic photosynthetic bacteria, using hydrogen sulfide to make carbohydrates without the production of free oxygen. Oxygen producing algae and cyanobacteria evolved about 2.8 billion years ago (Olson 2006). This free O_2 continued to be consumed, forming primarily iron oxides either magnate (Fe_3O_4) or hematite (Fe_2O_3), evidenced by large deposits of banded iron formations from 3 Ga to 1.8 Ga BP. Two things occurred to free oxygen into the atmosphere. For one, much of the Earth's crust was eventually covered over with the CO_2 forming carbonates. Secondly, the remaining exposed oxidative minerals were finally completely oxidized. After the oxidative consumption of primarily the free iron in the crust and oceans, a rather rapid increase in atmospheric O_2 occurred causing what is called the Great Oxidation Event (GOE). It is not known for certain what caused the sudden GOE. The presence of a reducing atmosphere for a period of time after oxygenic photosynthesis evolved is suspected. The most likely sequence is 1.) the evolution of the new oxygenic photosynthesis and beginning of production of large quantities of free O_2 2.) The consumption of most exposed iron and sulfur into oxides eliminating those reactions as an O_2 "sink". 3.) The rapid increase in Earth oxygen producing flora population.

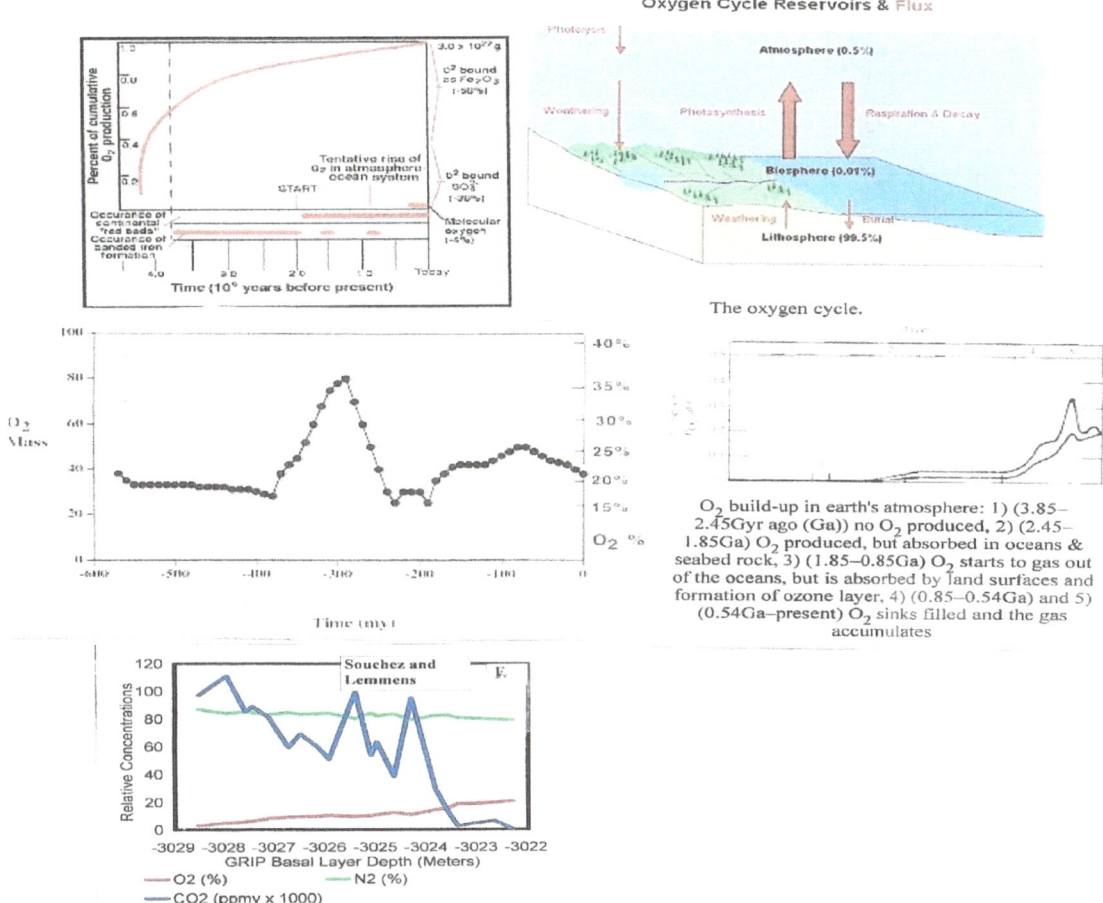

Figure 34, Oxygen evolution in Earth's atmosphere

(Upper left panel – Estimate of accumulation of oxygen and its disposition in the lithosphere, ocean and ultimately as stable atmospheric oxygen. Upper right panel – Graphic illustration of oxygen cycle. Middle left and right panels – Data for the atmospheric oxygen concentration from 550 million years BP estimated by Berner and Canfield (1989) showing the peak at about 300 Ma BP attributed to the rise of vascular land plants and widespread burial of organic matter in the vast coal swamps prevalent during that era. This episode is called the Great Oxygen Event. Bottom left panel - Estimate of CO_2 and O_2 from GRIP ice core basal layer samples.)

Figure 34 provides graphical sequences of atmospheric O_2 evolution. In the top left panel, we show the various phases of oxygen production on Earth. The first phase is the consumption of free O_2 by formation of terrestrial iron rich "red Beds" and banded iron formations that are found at many locations on Earth. At about 2.0 Ga Before Present (1 Ga = 1 Billion years), atmospheric free O_2 began to accumulate. During the period from 2.0 Ga to about 0.75 Ga there were two episodes where the banded iron formations resumed. As is shown in Table 1 for Carbon, we show a sketch of the oxygen cycling and reservoirs in the top right panel. In the top left panel, the total oxygen content on Earth is estimated to be 3.0 x 10^{22} grams of which 50% is bound as Fe_2O_3, 30% is bound as SO_3 and only 4% exists as free molecular O_2. The remaining 16% is bound in other oxidized

49

compounds. In the middle left panel, we present the Berner and Canfield (1989) computation of atmospheric O_2 concentrations back to 600 Ma BP, showing a large increase in O_2 levels during the Carboniferous and Permian periods from the rise in vascular land plants and the widespread burial of organic matter in the vast coal producing swamps. The middle left panel reflects the reduction in oxidative consumption of organic material reflecting the decrease in airborne CO_2 in the top left and middle left panels during this coal formation period. The Great Oxygen Event is shown. The bottom right panel shows these phases of oxygen evolution indicated in the caption. Finally in the bottom left panel, we have examined the Greenland GIRP ice core data of Souchez et al (1995) for samples from the bottom, basal region consisting of 6 meters of ice sheet base core material. What they found was an increase in CO_2 levels in the samples with increasing depth from the lowest level of 410 ppmv (nearly present day level) up to a level of 131,100 ppmv at the -3028.285 meters depth (the solid blue curve). Correspondingly, the IPCC measured O_2 level in the samples went from 20.8 % (the present day atmospheric level) to down to 3.38 % at the lowest sample (and earlier geological time). Diffusion and other gas effects have not been able to explain these data. We premise that the very high CO_2 level is indicative of the very high CO_2 levels at these early times on the order of 1 billion years BP and the corresponding O_2 level was at the early stage of O_2 accumulation.

Atmospheric Nitrogen and Argon

During the supernova explosion that created our nebula from which our Solar system was formed, the lighter elements above hydrogen and helium were more abundantly produced from the explosive fusion process. This of course included the lightest gaseous elements, after hydrogen and helium escaped, nitrogen and oxygen. Molecular nitrogen (N_2) is relatively non-reactive in an oxidative atmosphere and further has sufficient mass that it resists escape into outer space as did hydrogen and helium in early Earth. Therefore even though its relative supernova produced abundance is very low it has been retained in the atmospheres of Venus, Earth and Mars in significant quantities. On Venus it is the second most abundant atmospheric gas next to CO_2. On Earth it is the most abundant and on Mars also as the second most abundant behind CO_2. If CO_2 had not been scrubbed out on Earth by precipitation, Nitrogen too would be the second most abundant on Earth. Nitrogen has been identified as an essential element for life in the form of ammonia and nitrates. In the early atmosphere these are believed to have been produced mainly from fixation by lightning (Navarro-Gonzalez et al 2003). Argon, as would be expected as one of the non-reactive inert gasses, is also found on all three planets, since unlike the much lighter helium is too heavy to be lost to space.

Perhaps the most frequently used set of data for comparing past CO_2 levels and Global Temperature is the Berner and Kothavata CO_2 data shown in Figures 33 and 35 below and the temperature data compiled by Scotese in his PaleoMap Project. In a massive effort, Scotese (2003) has compiled global maps and estimates of climate conditions for 16 Earth time stages beginning 640 Ma BP when the Laurasia and Gondwanaland land masses were Earth's early continents. He then progresses through the plate tectonic continent movements, forming the super-continent Pangea and then the separation into the continents of today. As Figure 35, we show a correlation between Average Global Temperature and Atmospheric CO_2 over the past 600 million years (Hieb 2006).

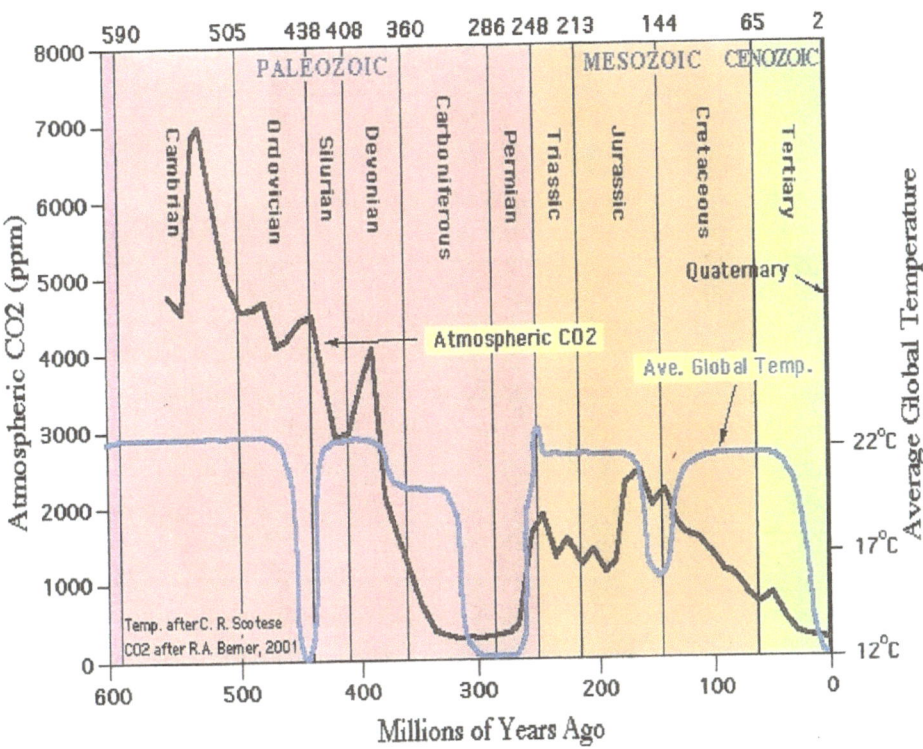

Figure 35, A composite graph of atmospheric CO2 and average global temperature from Hieb (2006) - obtained from Berner and Kothavaka (2001) and Scotese (2003) data

Figure 35 shows the CO_2 data from Berner and Kothavata (2001) and the Global Temperature from the PalioMap study of Scotese (2003). Note that he shows maximum temperatures of 22 °C whereas it is known that in the Paleozoic Era Earth temperatures were as high as 70 °C. His dips to 12 °C are for several known ice ages during that period. So we cannot consider his graph as quantitative. More recent deep sea core data have now provided definitive Earth temperature data going back to 500 million years BP (Lisiecki and Raymo 2005). These data are considered to be highly reliable and have high resolution correlation with Earth time from the 100,000 years astronomical cycles and Earth's magnetic reversals (known to good accuracy by radio-isotope dating of these events by many investigators).

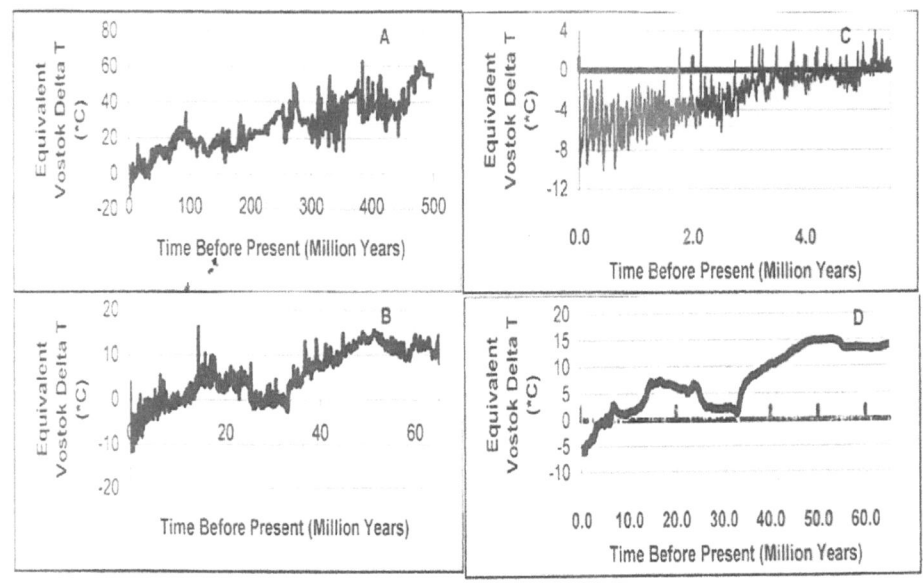

Figure 36, In three separate Panels, the Global Surface Air Temperature before present dating back to 500 Million years, compiled from benthic 18O proxy records from 57 globally distributed proxies (Lisiecki and Raymo 2005). Panel D is the 200,000 years average of Panel B showing a time averaged frigid Earth climate

We show these data for three scales in Panels 36A, 36B and 36C since we will refer to them in later sections. Panel 36D is smoothed data from Panel 36B. What are immediately noteworthy is that the presumed astronomical forcing cycles, causing the large climate fluctuations, changed from about a 41-43 kilo-year period, with then also a smaller temperature amplitude, to the present 100,000 year period with a much larger temperature amplitude beginning at about 1.1 Ma BP, seen in Panel C (shown in more detail in Figure 45 below). Before that, at about 2.7 Ma BP, the climate oscillations can be seen to be rather chaotic. The Earth was much warmer in the multi-million year BP past, with Antarctic glaciation first only occurring at about 35 Ma BP, but thawing at about 25 Ma BP before it permanently re-glaciated at about 14 Ma BP (Miller et al 2005). We will below, estimate the past Global temperatures and correlate between CO_2 and temperature relative to potential CO_2 greenhouse effects.

52

PART V — GLOBAL CLIMATE FACTORS - EARTH'S CARBON DIOXIDE AND OTHER CLIMATE FORCING PARAMETERS

In Part V containing Chapters 18 through 20, we will 1.) discuss the mechanisms that influence global carbon dioxide concentrations, 2.) discuss the mechanisms that influence global climate (temperature) including astronomical variations of Earths Solar intensity, 3.) examine any possible correlation between global CO_2 and global temperature and finally 4.) try to make some judgments regarding carbon dioxide and extraterrestrial global climate forcing and climate change in the future.

Chapter 18 – Carbon and Carbon Dioxide and Global Warming Mechanisms that Influence on Global Carbon Dioxide Levels

There are a number of factors that influence the overall global temporal behavior of carbon dioxide concentrations over the history of Earth. These factors are inter-connected, by both positive and negative feedbacks hence a variation of one will usually affect others. This feedback effect was first introduced by Bode with respect to electrical circuits (Bode 1945). A major factor is the fact that Earth has had liquid water. If however all the factors should remain constant for a period of time, they would all come into equilibrium and there would be an overall balance between them which would not be destabilized during the constant period. Without the introduction of some new influences during times of small changes, the factors are in quasi-equilibrium balance, as seems to have been the case in the present Holocene Era (post Ice Age present warm "interglacial" period) before humans began the current large use of fossil fuels and extensive cement production, releasing CO_2 back into the atmosphere. Before humans massive release of carbon dioxide, for the past million years Global CO_2 levels have remained within a very small range from about 180 to 300 ppmv. This compared to the much higher levels found by proxies in the past and shown in Figures 33 and 35. We can actually consider all of our primary atmosphere constituents as stabilized i.e. carbon dioxide, oxygen, nitrogen and argon. Some geo-physicists, most prominent are Dr. Hansen and Dr. Mann and of course the IPCC, consider the large increase in Global Carbon Dioxide levels, now as a de-stabilizing, catastrophic threat to our Global Climate. It is appropriate that the concern be addressed. We will show that the most dominant destabilizing influence on the Earth's atmosphere and hydrosphere and thus Earth's climate are extraterrestrial mechanisms that appear so massive as to be uncontrollable by humans and continually on a long-term basis, for the past million years, furnish the destabilizing force that causes the last ten severe Glacial-to-Interglacial-back to-Glacial cycles and we know will cause a new ice age, probably most certainly within the next 30,000 years.

Exchange Between Atmospheric CO_2 and Carbon Stored in Rocks, Carbonates – Effect of Weathering

The reason that Venus has retained its massive amounts of airborne carbon dioxide is due to its lack of liquid water since the global surface temperature has always been far above the vaporization temperature of H_2O and thus the absence of any precipitation on the planet. A major means (and the chronologically first means of removal of CO_2 from the Earth's atmosphere) was (and to some extent still is) from the precipitation weathering of calcium and magnesium bearing rocks by the formation of carbonates. The weathering process is the major component in the Berner and Kothavata modeling for CO_2 levels in the past 600,000 shown in Figures 33 and 35. Within terrestrial rocks, the Ca and Mg are primarily in the form of silicates. Continental weathering occurs to form, ultimately, Ca and Mg carbonates on the ocean floor (after transport of the weathering-derived carbonates to the sea by rivers) and deposition in sedimentary land regions. This weathering process has been accelerated in the past by exposure of new crust material in the up-lifting formation of mountain structures by plate tectonics. The weathering reactions involved are

$$CaSiO_3 + CO_2 \leftrightarrow CaCO_3 + SiO_2 \qquad (9)$$

$$MgSiO_3 + CO_2 \leftrightarrow MgCO_3 + SiO_2 \qquad (10)$$

The reverse reactions in Equations (9) and (10) represent thermal decomposition of carbonates primarily in the ocean and degassing of CO_2 to the ocean's surface, which fluctuates with ocean surface temperature (and the major cause of the large variation in Global Airborne CO_2 concentrations during the Glacial-to-Interglacial-back to-Glacial cycles). During periods of continental up-lift, it has been suggested that increased surface exposure of Ca and Mg bearing rock has caused several early ice ages. In particular, the 350 million year BP ice age (rise of Appalachian Mountains- Minkel 2006) and several little ice ages, 3.6 to 2.6 million years BP (rise of Himalayas – Leutwyler 2001).

Production and Destruction of Organic Material

CO_2 is released back to the atmosphere by oxidative weathering of organic matter resulting in degassing to the Earth's surface

$$O_2 \ + \ CH_2O \ \leftrightarrow \ H_2O \ + \ CO_2 \qquad (11)$$

In Equation (11) the arrows show a reverse reaction representing the result of photosynthesis in plant-life producing various forms of carbohydrates. Solar ultra-violet photons provide the energy for the photosynthesis reaction. A more exacting relation for photosynthesis is

$$6 \, CO_2 \ + \ 12 \, H_2O \ + photons \ \rightarrow \ C_6H_{12}O_6 \ + \ 6 \, O_2 \ + \ 6 \, H_2O \qquad (12)$$

The burning of fossil fuels (and forest fires) is represented by the Equation (11) reaction, which is self-sustaining when ignited. Decomposition of organic matter can occur in the presence of oxygen (aerobic) or without oxygen (anaerobic - fermentation). Both are produced by the breakdown of the material by living organisms. Most methane (CH_4) is produced with a small fraction of CO_2 in the anaerobic case. CO_2 is produced in the aerobic case. In global warming periods, airborne CO_2 and CH_4 levels will increase. Further, in the high latitudes as the frozen ground surface melts including frozen tundra, more CO_2 and NH_4 will be released to the atmosphere producing a positive feedback to Global Warming. A large amount of mobile CO_2 is dissolved in the oceans i.e. CO_2 reservoir – see Table 2 and Figure 35. It is estimated that the oceans store, in mobile form, 5 times the CO_2 content of the atmosphere at present airborne CO_2 levels. The solubility of CO_2 in water is inversely proportional to temperature, thus an increase in global temperature will increase the relative amount of airborne CO_2 to ocean dissolved CO_2. If extraterrestrial forces drive our Global Temperature to record breaking levels our mobile ocean pool could double or triple the airborne CO_2 levels because the ocean pool is so large. The Figure 37 provides a diagram of the pathways for the transfer of carbon bearing substances.

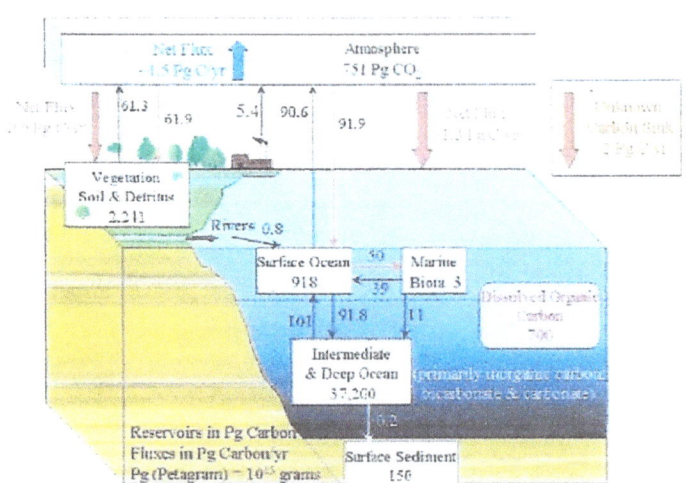

Figure 37, a diagram of the various global pathways for carbon dioxide transport which on stable climate conditions on large time scales, the Earth CO2 achieves an quasi-equilibrium balance. An estimate of build-up of atmospheric oxygen based on the basal layer Greenland ice core data (GRIP) of Souchez and Lemmens (1995) and estimates of Scotese (2003).

Figure 38, The large deposits of Calcium Carbonate comprising the, World War II popularized, "white cliffs of Dover"

These represent the carbon reservoirs on Earth given in Table 2. A very large fraction of the Earth's land mass and ocean bottom are covered with the 75,000,000 Gt of carbonates in Table 2. We show as Figure 38 the White Cliffs of Dover, England consisting of the calcium carbonate chalk from shells of marine microorganisms (shown in insert).

Figure 39, The Grand Canyon layered deposits of carbonates

The Grand Canyon is another readily observable example of the large amount of Earth's crust containing carbonates. At some points the canyon erosion is to a depth of over one mile, displaying expansive layers of limestone and shale deposits as seen in Figure 39.

Figure 40, Caverns frequently found in deep limestone (Calcium Carbonate) deposits

In the southwest and eastern United States, the large deposits of limestone formed during the Cambrian Period measure from 7,000 to 11,000 feet in thickness and is evidenced by numerous caverns such as seen in Figure 40.

Figure 41, Weather resistant dormant volcano core (plug) called Shiprock in New Mexico. Surrounding debris is non-core weather eroded volcano material

After the solidification of the Earth's crust, there are records of intensive volcanic activity as evidenced in the U. S. Southwest of ancient volcano cores such as shown in Figure 41. The Pacific Ocean bottom is littered with tens of thousands of un-eroded extinct volcanoes. Intense volcanic episodes are proposed by Kopp et al (2005) and others to have occurred to break the Earth out of the "Snowball" Earth grip and other early ice ages, this creating our second stage CO_2 atmosphere.

Of course, major on-going volcanic activity is still directly increasing the global concentration of CO_2 as shown in Table 1, but at a much reduced rate.

57

Photosynthesis and Transitions in Its Evolution

Early algae and bacteria were in various forms including single-cell organisms with most of these species still present on Earth. Large deposits of these are found on the Northern coast of Canada. Although they are not as complex an organism as land plants, the biochemical process of photosynthesis is the same. Early bacteria were an oxygenic, using various molecules such as hydrogen and sulfur compounds as electron donors. Land based life, such as cyanbacteria, evolved to utilize the energy of light photons for their energy source and oxygenic photosynthesis through the carboxylation of CO_2 to form phosphoglyceric acid, a 3 carbon acid. This is called the C3 photosynthesis pathway or Calvin cycle. Nearly all trees, most shrubs, herbs and forbs, and cool-season grasses and sedges use the C3 pathway.

About 40 million years ago, attributed to the drastic decline in atmospheric CO_2 (Osborne and Beerling 2005, Cerling et al 1993) to its present historically very low level (see Figures 33 and 35), a more efficient mechanism for photosynthesis, the C4 pathway, evolved. The process utilizing 4-carbon oxaloacetic acid made more efficient use of Earth's declining airborne CO_2 levels. Presently over 8,000 species of plant life have made C4 adaptations.

Chapter 19 – Environmental & Astronomical Cycles Affecting Earth's Climate

Sun Spots and Total Solar Irradiance

There are a number of factors that affect the global temperature. The most direct is, of course, the Suns un-attenuated radiation intensity quantified by the Total Solar Irradiance (TSI), measured in units of Watts per meter^2. TSI has been monitored, at the top of the Earth's atmosphere, by satellites since 1978 and currently there are two satellites collecting TSI data (Wilson and Mordvirnov 2003). The Sun's intensity has been found to vary with the Sun's sunspot frequencies over this same period. Figure 42 provides an image of the Sun at a period when the sunspot intensity was at its low period. Figure 43 provides the TSI data from 1700 to 2005 and the sunspot number from 1750 to 2005.

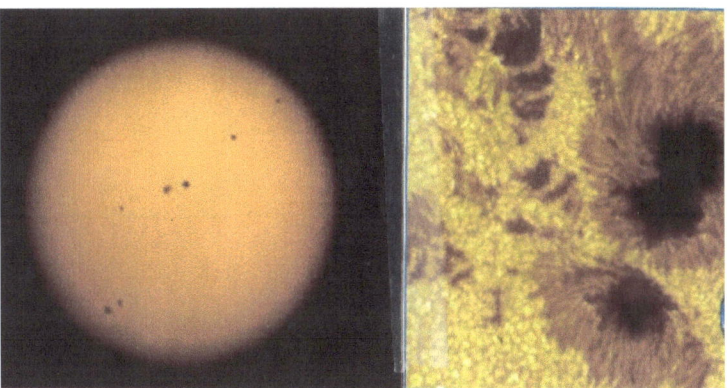

Figure 42, Left panel – Image of Sun during minimal sunspot intensity. Right panel – Enlargement of several sunspots showing structure from magnetic field variation

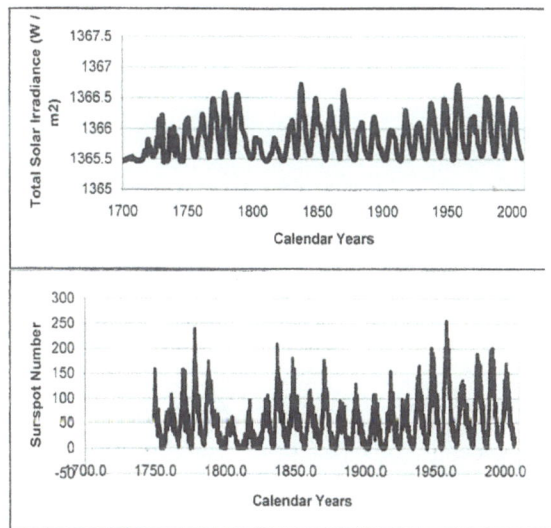

Figure 43, Upper Panel – NOAA published data of Wilson and Mordvirnov (2003) of Total Solar Irradiance (TSI). Lower Panel – NOAA published Sunspot Number data. What is shown is close correlation between the observed variation in Sunspots and the observed variation in the Solar intensity above the top of the atmosphere by satellite monitoring.

59

It has been known that the Suns intensity varies directly with sunspot levels caused by magnetic fluctuations on the Suns surface and interior. It has been observed that the greater the sunspot activity the more intense the unattenuated Solar Irradiation at the top of Earth's atmosphere. Sunspots have been monitored visually since Galileo's invention of the telescope and a correlation was made between a significant reduction in sunspot number and the Little Ice Age (Maunder Minimum) from 1600 to 1700 AD. From Figure 43, it is clear that Solar Sunspot intensity affects the Solar irradiance energy deposited at the top of the Earth's atmosphere – but the range is small. The NOAA National Geophysical Data in Figure 43 shows the sunspot cycle frequency to be about 11.1 years. As seen in Figure 43, the Sun was at its 11 year minimum intensity in 2009. From observation of the evolution of other stars (our Sun is a medium sized star) we know that our Sun's intensity is gradually increasing but at a very slow rate. As we noted in early Earth, the Sun's intensity was about 80% of our present level until the fusion furnace gained full strength.

The Milankovitch Astronomical Cycles

There are also cycles related to our Earth's orbit, which we have mentioned, called the Milankovitch astronomical cycles. Currently astronomers are aware of three orbital Earth motion cycles shown in Figure 44 and Figure 45 that affect the Earth's Solar intensity.

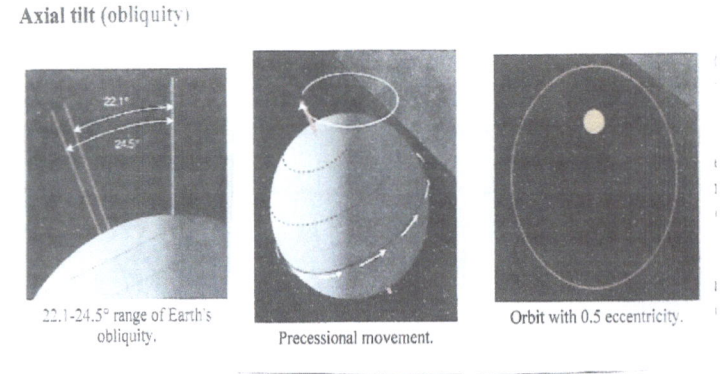

Figure 44, Illustration of Astronomical (Milankovitch) Cycles affecting Earth's Solar Intensity.

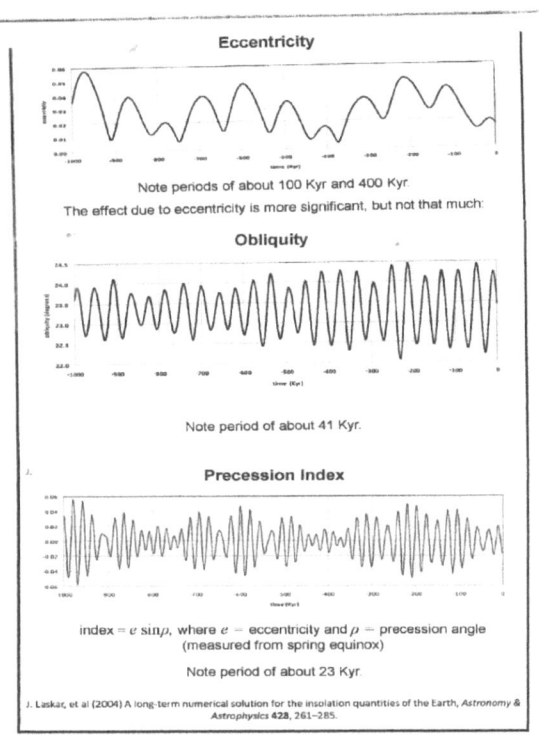

Figure 45, The Oscillatory Astronomical Cycles for Eccentricity, Obliquity and Precession Influencing Solar Intensity at the Top of Earth's Atmosphere Bach to 1.0 Ma BP.

The shape of the Earth's orbit around the Sun is not a perfect circle but a variable ellipse that changes in eccentricity from -0.03 to +0.02 (presently +0.017) with approximately 100 kilo-year cycles. Also, the tilt of the Earth with respect to the plane of orbit around the Sun (ecliptic inclination) changes from 22.1° to 24.5° (presently 23.44°) with a 41-43 kilo-year cycle. Finally, the Earth's axis precesses (as like the wobbling of a spinning top) with a 26 kilo-year cycle (It has been known for many years that the Earth's equinoxes have been pressessing at the annual rate of 50.27 seconds of arc.). These variations change the Solar intensity over large time frames and is now believed by the majority of geophysicists to cause the Glacial-to-Interglacial-back to-Glacial climate cycles.

Most geophysicists now accept the fact that the glacial terminus periods during at least the last 11 major glaciations and large climate fluctuations shown in Figure 47 are primarily driven by the coupling of the three astronomical (Milankovitch 1998) forces i.e. 1.) variation in the tilt of the Earth towards to Sun, 2.) variation in the elliptic of its orbit and 3.) variation in the Earths orbital eccentricity with Earth having cycle frequencies of about 21, 41 and 100 kyrs, respectively. These motions provide the present 100,000 year cycles shown in Figure 47. Before the present 100,000 cycles, Raymo and Nisancioglu (2003) show evidence of 41,000 year cycles. Figure 36, Panel C shows the climate cycles transitions. Over decades many correlations have been attempted between these three forcing cycles with only limited success [i.e. Raymo and Nisancioglu, 2003, Wunsch, 2004, Rial and Anaclerio, 2000, Tziperman et al, 2006, Berger et al 2006]. In Figure 46, we illustrate the premised harmonic effect of the three 100,000, 41,000 and 21,000 astronomical cycles. In Figure 46, trigonometry sine wave functions were used and the magnitudes of the three cycles are set equal, which is not to be true for the Milankovich hypothesis applied to Global Warming.

Example: 100, 41, and 21 Cycle Years
Harmonic Amplitudes

Figure 46, *Illustration of Cyclic Harmonics Producing In-Phase Resonance and Out of Phase Destruction*

It is found, primarily from Antarctica and Greenland ice cores, that Global Surface Air Temperature, Global Carbon Dioxide Concentrations and other observable Earth parameters currently vary on approximately a 100,000 year period of climate change in agreement with the Milankovitch astronomical theory. We will discuss these cycles in detail. These are long-term changes and, of course, our TSI program has not had time to observe them. The EPICA Antarctica site ice core data, by proxy, do reflect these cycles as shown in the Figure 47, clearly showing the 100,000 years cycles for both Global Temperature and airborne CO_2.

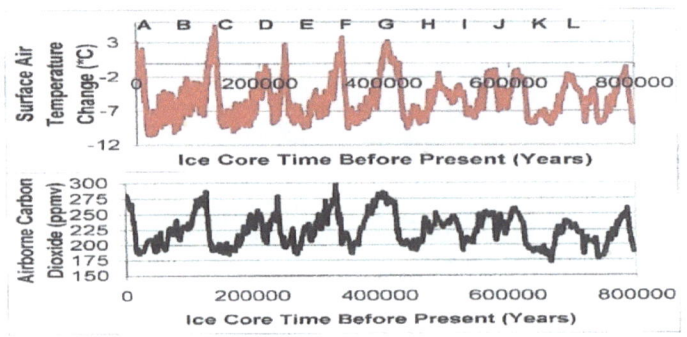

Figure 47, *The Global Surface Air Temperature and Global Airborne Carbon Dioxide Concentrations over the past 800,000 years before present based on the EPICA site Antarctica ice core data. The A through L are our method of labelling the interglacial peaks, recognizing some at greater than 500,000 BP as reflecting the 40,000 year cycles*

Above as Figure 47 we have provided the NOAA published measured Global CO_2 levels and Global air surface temperature data from the Antarctica EPICA ice core data (Jouzel et al 2007, Luthi et al 2008), demonstrating the correlation between the two. The 100,000 year cycle is evident as the presently dominant cycle in Global climate but the 41,000 and 21,000 year cycles are seen to be present in the earlier ice ages in Figure 47 and even more so at earlier times BP shown in Figure 36. We see that we now have ice core (Antarctica, Greenland, glaciers), ocean core and land based core evidence of CO_2, methane, air temperature, etc. levels dating back to several million years that confirm the 100,000 year current cycle. We show more completely in Figure 47, the EPICA ice core data with records of both CO_2 and temperature going back 800,000 years BP. We will use the EPICA data in our analysis below of 12 Glacial-to-Interglacial-back to-Glacial cycles, which we designate as Interglacial (IG) cycles A through L (For the period from 1 Ma BP to 800 ka BP we consider 8 cycles).

Other Behaviors Influencing Climate – Dansgaard-Oeschger and Heinrich Events

Of lesser importance, scientists have also observed a cyclic oscillation in global climate with a period of about 1470 years, called Dansgaard-Oeschger (D-O) events. These events have been mainly seen in the Greenland ice cores, thus a Northern hemisphere effect, and have not been regular until the last ice age. The behavior is a rapid warming of temperature (up t0 8 °C in 40 years), followed by a cool period lasting a few hundred years (Bond et al 1999). This cold period resulted in an expansion of the polar front, with ice floating further South across the North Atlantic. Figure 48 provides data from the Greenland GRIP and NGRIP ice cores showing the D-O oscillations from 50,000 to 27,000 years BP during the last ice age. Yang and Rial (2006) have attributed the D-O events to internal oscillations of the thermohaline ocean circulation (which we will discuss in later sections), and in particular to intermittent termination of the Nordic and Labrador Sea currents. The fact that the D-O type events are infrequent during the warm interglacial periods is believed to be due to the 100,000 years cycle which modulates the thermohaline oscillations during the Interglacial Period.

Kobashi et al (2008) suggest that the abrupt warming event at about 11.3 kilo-years BP was a smaller manifestation of the D-O events. The more recent Little Ice Age (-400 to -200 years ago) is suspected to be, perhaps, the cold part of a D-O cycle, which would explain our current warm period, without justification by anthropogenic global warming. A second behavior observed during the last glacial period is what is termed the Heinrich events. The primary observable effect has been a cold spell sometimes preceding a D-O event during which armadas of icebergs broke off from glaciers and the Greenland ice cap and floated South in the North Atlantic. Evidence of this is the observed melting of icebergs dropping glacial rock onto the North Atlantic sea floor as "ice rafted debris". Again there is no conclusive explanation for these Heinrich events either or that they have always preceded the D-O events. Between the D-O and Heinrich events, the last ice age exhibited a large number of small temperature oscillations making it a very noisy climate period as seen from Figure 48.

Figure 48, Examples of other cyclic oscillations in Global climate, of a lesser magnitude than the

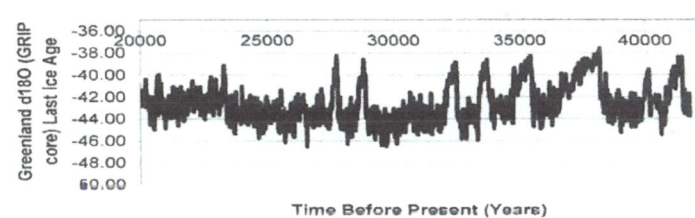

dominant 100,000 year astronomical cycles. Shown here are the last glacial period oscillations from Dansgaard-Oeschger events and Heinrich events.

We have to consider these oscillations minor since all the while most of Canada, Northern Europe and Northern Russia were covered with ice sheets thousands of meters thick.

Chapter 20 - Present Day Data Collection of Global Climate Parameters

An Earth-bound effect on the Solar intensity of the Sun's rays, in terms of heat energy absorbed on the surface, is the concentration level of airborne carbon dioxide, methane and nitric oxide in our Earth's atmosphere. Earth bound global CO_2 concentrations have been measured at terrestrial stations since 1957. NOAA also measures Sea Surface Temperatures (SST) at various stations around the globe. As Figure 49, we provide the directly measured SSTs.

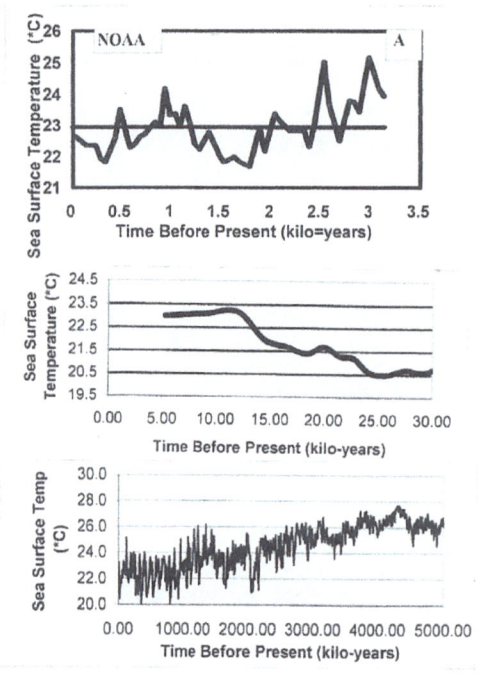

Figure 49, Three panels of NOAA Sea Surface Temperature data. Upper Panel – Sea Surface Temperature variation over the last 3,000 years in our present Interglacial warming (Plateau Region) period. Middle Panel – The Sea Surface Temperature variation during the termination period (Rise Region) from the past ice age. Bottom Panel – Sea Surface Temperature from ocean bottom cores back to 5 million years BP, showing the large oscillations from the 100,000 year cycles from 1.1 million years BP and a trend for decreasing global temperature back to 5 million years BP.

Finally, NOAA provides Sea Level data which is a direct measure of glacial and ice cap melting. We show the NOAA data for Sea Level variation as Figure 50.

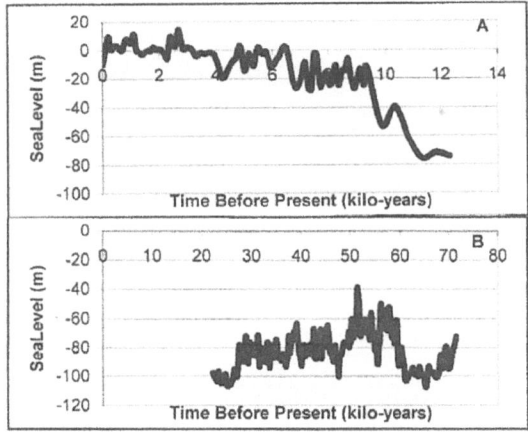

Figure 50, Two panels of NOAA Sea Level data. Upper Panel – Sea level rise beginning at 12,000 years BP at the termination of the last Ice age and a gradual increase as glaciers, the Greenland ice cap and Antarctica are continuing to melt from gradual climate warming. Lower panel – Sea level consistently about 80 meters below present level during last ice age.

There are also numerous less direct proxy data providing measures of the various factors mentioned above relative to global air temperature and global airborne carbon dioxide concentrations. Some of these will be examined in the following section. Data has been collected on Average Global Temperatures (AGT) now going back over a century i.e. back to 1880. These data are based on daily temperature recordings from weather stations world-wide. Dr. James Hansen (Hansen et al 2007) has been a pioneer in the program to monitor how the Earth is doing with respect to global temperature change. His data reports that the calendar year 1998 currently holds the record for the highest global temperature recorded since the beginning of land based, meteorological stations data, in 1880, with 2005 taking second place. A controversy exists relative to the influence of man-made greenhouse gasses (GHG) in the relatively steady increase in global temperature. This has been subject of the Intergovernmental Panel on Climate Change. Opponents to the GHG theory are scientist that argue that cyclic variation in solar intensity, quantified by the Total Solar Irradiance (TSI), is the major factor causing the current relatively steady mono-tonic by increasing AGT change, in other words extraterrestrial forcing. We shall in the following sections offer substantiated scientific evidence relative to various possible "forcing" influences on AGT including airborne carbon dioxide and other GHGs.

There is evidence of a "recent" consistent decrease in Earth temperature based on the 500 million year data stacking (Lisiecki and Raymo 2005) and Sea Surface Temperature decrease. Figure 36, for Surface Air Temperature, and Figure 49, for Sea Surface Temperature, shows a nearly monotonic decrease in Global Temperature going back 500 million years BP. Because the oscillatory cycle frequency for Air and Sea Temperature undergoes a transition from a 41,000-43,000 period to the present 100,000 year period, it is most probable that the temperature decrease is due to astronomical sources. But there are other considerations such as the plate tectonics reconfiguration of the continents and ocean circulations. The possibility exists however that the higher atmospheric CO_2 levels could impose a Global Warming influence. This will be extensively examined, in particular examining past Interglacial periods.

Part VI – ATTEMPTS TO CORRELATE LAG and LEAD TIMES FOR CARBON DIOXIDE AND GLOBAL TEMPERATURE FROM ICE CORE DATA

In Part VI containing Chapters 21-25, we report others work in attempting to correlate global carbon dioxide concentrations with global surface air temperature and support or refute IPCC's position that CO_2 is the cause of Global Warming.

The major problem, in evaluating global climate, is how to directly compare Earth carbon dioxide and Earth temperature. Many geo-physicists have searched for ways to quantitatively evaluate the correlation between Global Airborne CO_2 levels and Global Surface Air Temperature. To do this we herein develop a correlation model to provide an equivalence between Surface Air Temperature in °C and CO_2 levels in ppmv, such that a direct quantitative comparison may be made.

A significant issue is the basic question "Does human carbon dioxide production affect global temperature and thus our global climate?" The most notable comparisons between airborne CO_2 levels and global surface air temperatures has been that of Vice President Gore (Gore 1992, 2006) with his two texts and his motion picture. He has consistently presented the ice core data such as our Figure 47 in his Global Warming discussions with the premise that Global Atmospheric Carbon Dioxide drives Global Temperature at the present levels of airborne carbon dioxide.

We have shown that initially, after the early Earth atmosphere of hydrogen and helium was dissipated by their escape into space, primarily assisted by the Solar wind, the Earth accumulated a carbon dioxide atmosphere similar to the present Venus atmosphere. This was concluded by computing the carbon dioxide airborne composition that would exist if all the Earths present pooled carbon given in Table 2 were present as airborne CO_2. That calculation showed that an Earth atmosphere primarily of about 98% CO_2 could have been present except for precipitation "scrubbing" out as accumulation occurred. Additionally, small amounts of Nitrogen and Argon and also very small amounts of O_2 since most of the initially free O_2 would react with Earth metals to form oxides. The CO_2 (and the water forming the initial oceans) would have been from the very active volcanoes of that early period (see Table 1). Before life began, as shown in Figures 33 and 35 and modeled by Berner and Kothavala, a significant fraction of the initial 98% CO_2 was swept out of the atmosphere in rainwater and quickly reacted with Earth minerals, primarily Ca and Mg to form vast land and ocean deposits of $CaCO_3$, $MgCaCO_3$ and $MgCO_3$, in many locations many kilometers thick. Realistically, we must know that at no time did Earth have a 98% CO_2 atmosphere because, if so, Earth could have been locked into the same condition as Venus with such high surface temperature (estimate 250 **°C**) that there would be no liquid water, no precipitation and no scrubbing out of the CO_2. Then as atmospheric CO_2 was accumulated from volcanic activity it was simultaneously scrubbed out. There are some data on CO_2 and temperature levels early in Earth's history. Life first began with CO_2 consuming ocean phytoplankton and land algae, that further consumed the airborne CO_2 and began producing airborne oxygen, although some oxygen is known to have been present in the earliest atmospheres (Walton 1976,). The early O_2 was consumed by oxidation of oxidizing metals, predominantly iron which is evidenced by the large amount of red soils in the Earth's crust. Only after the crust oxidation was complete did O_2 begin to accumulate in significant quantities in the atmosphere. With the accumulation of oxygen, oxygen consuming life species evolved first with plankton and eventually coral and other crustaceans in the oceans. This process- further consumed CO_2 to form $CaCO_3$ shells.

Evidence of a Lag of Carbon Dioxide Temporal Behavior With Respect to Global Temperature

In examining in detail the ice core data for Global Carbon Dioxide Concentrations and Global Surface Air Temperature, several investigators have noted an off-set between the abrupt rise portions (called terminal regions) of the glacial to interglacial transition. Indermuhle et al (2000) compared the Antarctica Taylor Dome ice core data to the Antarctica Vostok site ice core data, as shown in Figure 51, and noted in four Antarctic warming events, a time lag of the CO_2 interglacial rise with respect to the rise in temperature of 900 years ($R^2 = 0.83$ for the shifted time scales).

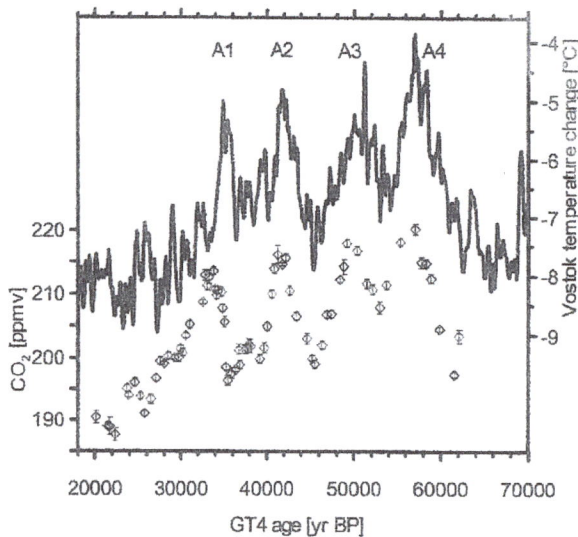

Figure 51, Comparison of Carbon Dioxide and Air Temperature Data from the Ice Cores Showing a Time Lag for Carbon Dioxide.

Hansen et al (2007), using the Vostok CO_2 and temperature data, did a similar scale shifting comparison and found a correlation to 92.5% CL for a shift of 700 years. We show their Figure 1 as our Figure 52 herein.

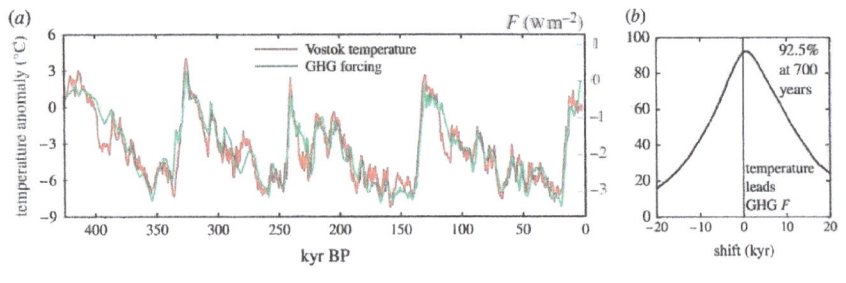

Figure 52, Comparison of Carbon Dioxide and Air Temperature Vostok Ice Core Data by Hansen et al (2007) Showing Lag of Carbon Dioxide.

Soon (2007) also examined the Vostok ice core CO_2 and temperature data for the present and past three interglacial periods dating back 420,000 years BP. Their comparisons concluded that Antarctic warming tends to lead the rise in CO_2 concentrations by several hundred years. We show his Figure 1 as our Figure 53.

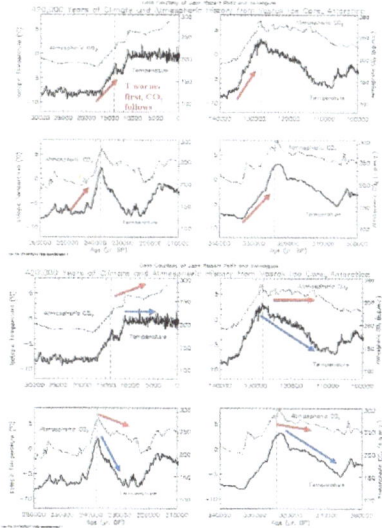

Figure 53, Soon (2007) Comparison of Vostok Ice Core Data for Last Four Interglacial Periods Showing Carbon Dioxide Lagging Atmospheric Temperature.

Most recently Stott et al (2007) found that deep-ocean temperatures lead the rise in atmospheric CO_2 between 19 and 17 thousand years BP as the Earth was emerging from its last ice age. If CO_2 changes indeed lag global temperature changes, then global warming cannot be induced by the greenhouse effect of atmospheric CO_2 but, as premised by many investigators, by extraterrestrial causes such as astronomical influence. The primary difficulty in making the CO_2 and temperature comparisons is that it is awkward comparing the CO_2 data in ppmv and temperature in °C units. In the next sections, we use a scaling transformation to more accurately examine the time sequence correlation between the CO_2 and temperature data of the Antarctic EPICA site.

A Problem with the Indermuhle et al (2000), Hansen et al (2007), Soon (2007) and Stott et al (2007) Lag Results - The Gas Age and Ice Age Differences in Ice Cores

Above we have presented the data of Indermuhle et al [2000], Hansen et al [2007], Soon [2007] and Stott et al [2007] indicating a lag in changing CO_2 levels behind increasing global temperature. We know from careful review of those writings that due to gaseous diffusion of the CO_2 in the ice core samples there is some difference between what is called the ice core determined gas (CO_2) age and the ice temperature assessed age. There has been considerable uncertainty as to the magnitude of this gas age to ice age difference, Δage, in evaluating ice core data for the temporal scale for past CO_2 concentrations relative to air temperature (Barnola et al, 1991). These uncertainties arise in estimating the beginning of the Terminus Rise Region (glacial terminations) and the question of whether Global Temperature changes initiates CO_2 change or the reverse happens.

Here in our work, there are several significant features of our data analysis that strongly support the premise that extraterrestrial forces are initiating the Global Temperature changes and the CO_2 changes occur as a homeostatic reaction primarily due to the inverse temperature dependence of the CO_2 ocean water solubility. We have found that Global CO_2 increase lags Global Temperature increase in the initial Terminus Rise Region of the Interglacial cycles. We find that the slope of the Rise Region for the CO_2 is considerably less than the slope of the temperature Rise Region, which we show would be true from a homeostatic view-point considering a lag for the warming of the oceans (based on Sea Surface Temperature data). We find that after the Rise Region is completed, the CO_2 continues to increase in the Interglacial Plateau Region where in some cases the Global Temperature on an average slightly decreases. This would be true as the Sea Surface Temperature is continuing to warm and expelling CO_2 into the atmosphere. We find that at the end of the Interglacial Plateau Region, when the Global Temperature begins to return to the glacial levels, the CO_2 continues to increase. We find a mean lag time in the Rise Region of 1391 years. This mean lag time is further increased through the Interglacial Plateau Region by another 1230 years making a total lag at the end of the Plateau Region of 2521 years. If the lag were simply a difference in accounting for the gas to ice age, Δage, the lag would not further increase in the Plateau Region. We have also shown in later sections, with the Pagani et al (2005, 2006) data, that Global Temperatures are very loosely coupled if at all for a wide range of past temperatures and carbon dioxide levels.

Chapter 22 - IPCC Attempt to Directly Correlate Between Atmospheric CO2 and Mean Global Surface Air Temperature

The International group IPCC has developed an analytical method of assessment for the correlation between atmospheric CO_2 concentrations and mean Global Surface Air Temperature (in other words global climate). This analytical method involves the sensitivity of climate to change in the radiative forcing parameter defined by IPCC as the energy deposited on the Earth's surface in units of Watts per square meter of Earth's surface. It is well accepted that there are both positive and negative feedbacks to any change in CO_2 levels affecting change in global temperature and positive and negative feedbacks to changes in global temperature affecting CO_2 levels. For example as we have mentioned several times, the solubility of CO_2 in water decreases with increasing water temperature, thus if increased CO_2 results in increased radiative forcing and increase in temperature by decreasing the Earths albedo (albedo is the fraction of energy reflected from the Earth's surface) then additional CO_2 is released from the oceans from increased ocean temperature. Similarly, if astronomical effects increase the Earth's temperature, ice sheets in the northern hemisphere will melt, increasing the exposed darker land mass and thus decreasing the albedo and further increasing the radiative forcing parameter and Global Air Temperature. Both of these are positive feedback effects tending to accelerate the changes. The major temperature feedback is from water vapor in the atmosphere. As the atmosphere warms in response to a forcing, the carrying capacity of the space occupied by the atmosphere for water vapor increases nearly exponentially in accordance with the Clausius-Clapeyron relation. Since water vapor is the most important greenhouse gas next to CO_2, the growth in its concentration caused by atmospheric warming exerts an additional positive forcing, causing temperature to further rise. IPCC has considered approximately 20 temperature feedbacks, most of them with little overall effect on global warming. To analytically model these feedbacks in earlier reports, IPCC has applied the Bode feedback equation developed by Bode (1945) for electronic circuits. For small changes exerted by feedbacks the Bode equation approximates linear behavior.

IPCC Bode (1965) Type Climate Feedback and 2007 Fourth Assessment Report (FAR) Estimate

IPCC's method of evaluating climate sensitivity involves the use of several carefully defined variables. They are - an absolute land and sea surface temperature, T_S, and an equilibrium temperature change parameter, $\Delta T\lambda$, in response to all anthropogenic-era radiative forcings and consequent temperature feedbacks i.e. further changes in T_S that occur because T_S had already changed in response to a forcing arising in response to the doubling of pre-industrial CO_2 concentration. This involves three factors, radiative forcing, ΔF; a no-feedback climate sensitivity parameter, k, and a feedback multiplier, f. The change then is given by

$$\Delta T\lambda = \Delta F2x \text{ k f where } f = (1 - bk)\text{-}1 \text{ and b is the sum of all climate related} \quad (13)$$
$$\text{feedbacks.}$$

If b is positive then f is less than unity and if it is negative then f is greater than unity. The 2x suffix denotes for doubling. IPCC has estimated a number of feedback factors. In IPCC (2007) they give a value of 3.405 W m^{-2} for ΔF_{2x} anthropogenic-era forcing and values of k = 0.313 °K W^{-1} m^2 and b = 2.16 W m^{-2} °K^{-1}. Then the CO_2 doubling temperature is estimated by IPCC to be

$$\Delta T_\lambda = \Delta F2x\ k\ f\ =\ 3.405 \times 0.313 \times 3.077\ =\ 3.28\ °K \text{ per doubling of CO2.} \qquad (14)$$

This is termed the climate sensitivity temperature rate. IPCC provides a range from 2 to 4.5 ° K with a 66% probability. When other feedbacks are considered the calculations are more complicated but straightforward.

Here, in the analysis of the past 800,000 years BP interglacial warm cycles, we correlate Earth's atmospheric temperatures with Earth's global carbon dioxide concentrations. We will show that there are several significant features of our data analysis that strongly supports the premise that extraterrestrial forces are initiating the Global Temperature changes, even in the warm Interglacial plateau, and the large pre-industrial CO_2 changes have occurred as a homeostatic reaction primarily due to the inverse CO_2 solubility of the oceans to temperature change.

Chapter 23 – Proxy Data to Reconstruct Paleoclimate Conditions of Global Carbon Dioxide Concentrations and Global Surface Air Temperature

A number of different techniques have been developed to estimate climate variations over a range of time periods from years to millions of years. Table 3 provides a list of these.

Table 3, Primary Sources of Proxy Data for Paleoclimate Estimates

---

Glaciological (Ice Cores)
Biological (Tree Rings - density, width, isotopes; Pollen; Insects)
Geological (Marine sediments - ocean sediment cores;
       Organic sediments - benthic and planktonic;
       Inorganic sediments)
Historic (Meteorological, environmental, biological)

Meteorological, ice core and tree ring data are the primary proxy data that have been useful for estimating our climate in our present interglacial (Holocene) period and will be presented in the following sections. Ice core and marine sediment (cores) are shown to be useful for dating back to 600 million years ago (see Figure 49).

Ice Core Dating Method

There are 19 Ice Core Stations in Greenland, 12 in Antarctica and 6 others located at other around the world sites such as glacials. The major ones for which NOAA has published extensive data are EPICA and Vostok in Antarctica and GRIP/GISP and NGRIP in Greenland.

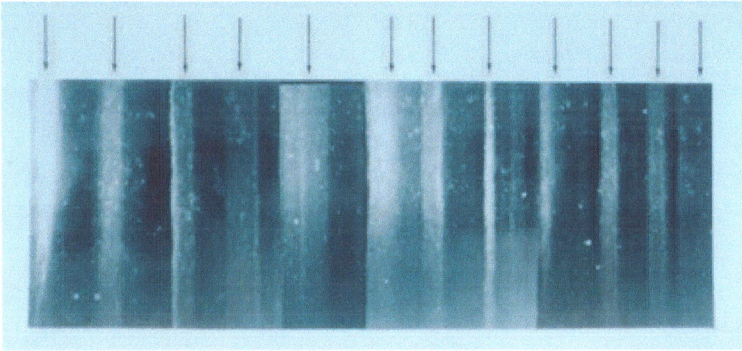

Figure 54, GISP 2 Ice Core Sample 19 cm at 1855 meter depth.

Figure 54 provides a photograph of a 19 cm long ice core covering 11 seasons from Greenland's GISP 2 site. Each light region represents the summer season and the dark regions represent the winter season. This core section was taken at a depth in the ice of 1855 meters. Calibration is necessary since the years per meter increases with depth from compaction i.e. at 800 meters the chronology is 5000 years per meter.

73

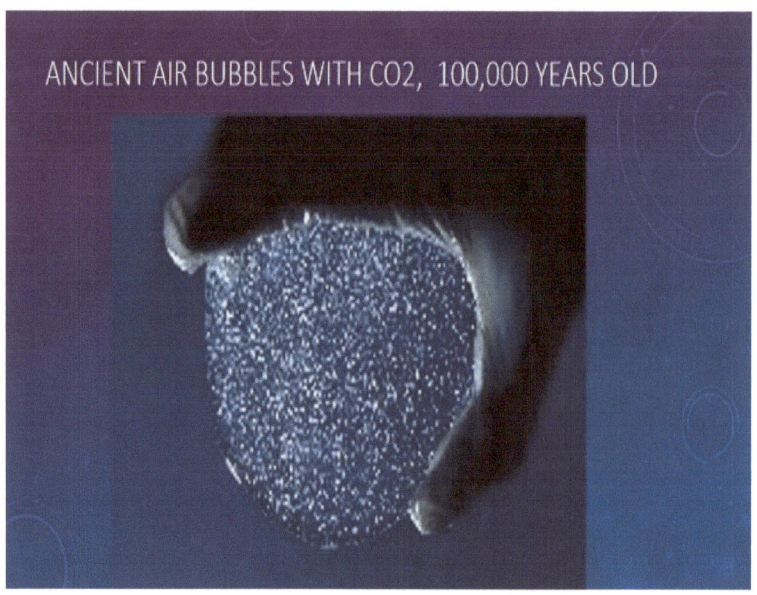
ANCIENT AIR BUBBLES WITH CO2, 100,000 YEARS OLD

Figure 55, Cross section of Ice Core showing tiny air bubbles containing ancient CO2 and temperature information

Figure 55 shows a cross section of an Ice Core containing tiny air bubbles able to provide CO_2 and other airborne constituents. The temperature dependence of the ice core samples is derived from the variation the heavy oxygen isotope ^{18}O relation to ^{16}O in the ice water molecules. The lower vapor pressure of the $H_2^{18}O$ means that it passes more readily into liquid state than water vapor made up of the lighter oxygen isotope (Dansgaard 1961). This temperature dependence of the two isotopes enables the relative temperatures when the ice was formed to be determined. Carbon dioxide concentration variation is determined at the time the ice was formed is determined from the air bubbles trapped in the ice cores. A proxy is the use of something other than the quantity at issue to represent this other quantity at issue. So technically ice core data for CO_2 is a direct measure of the CO_2 in the tiny bubbles at the time in the past that the bubble was formed – not a proxy. Variations in other constituents can be determined in the ice i.e. atmospheric aerosols, calcium, silicon, aluminum and dust concentrations (indicating a dry climate). Vice President Al Gore has used the ice core data on numerous occasions to promote his Global Warming hypotheses including the description of the cold spell referred to as the Younger Dyras Event. But his primary use is to maintain that carbon dioxide drives Global Warming.

Tree Rings Dating Method

Figure 56 provides a photograph of a horizontal cross section of a tree trunk covering many years of tree growth.

Figure 56, Cross Section of Tree Trunk, Annual Growth Used for Climate Dating.

The study of annual growth of tree rings to obtain climate data is called dendroclimatology. A number of factors influence the rate of growth of a tree and hence the internal growth of the annual outer-most tree ring in the trunk of the tree. If the other factors can be assessed, then the environmental climate to which the tree was exposed during that year can be estimated. Hence, the warmer the climate for that season, the larger the tree ring and vice-versa the larger the tree ring the warmer the season. The use of dendroclimatology has been successful in archeology in conjunction with radio-carbon dating. A number of research groups have applied the science to estimating past temperatures in our present Holocene Period with respect to Global Warming and the effect of Green-house Gasses. IPCC has used a significant amount of these data in their 2001 and 2007 reports. The work of Dr. Michael Mann and his colleges (Mann 1999) and data from the Yamai tree has produced the Hockey Stick shaped graph used by IPCC. Several scientist have performed an audit of this tree ring analysis and found that some of the data was falsified and fabricated, perhaps now putting more credence in ice core data. This will be discussed in the next section.

The Tree Ring Controversy

As noted above, a research group directed by Dr. Michael Mann (1999) published peer reviewed data showing a large increase in Global Temperatures in conjunction with the large increase in man-made carbon dioxide emissions. These data were the impetus for IPCC to forecast very large future increases in Global Temperatures and that we were nearing an irreversible runaway climate. As noted in the Introduction, Congress drafted a Cap and Trade legislation and the White House report stated "Warming of the climate is unequivocal, and man-made gases are primarily to blame."

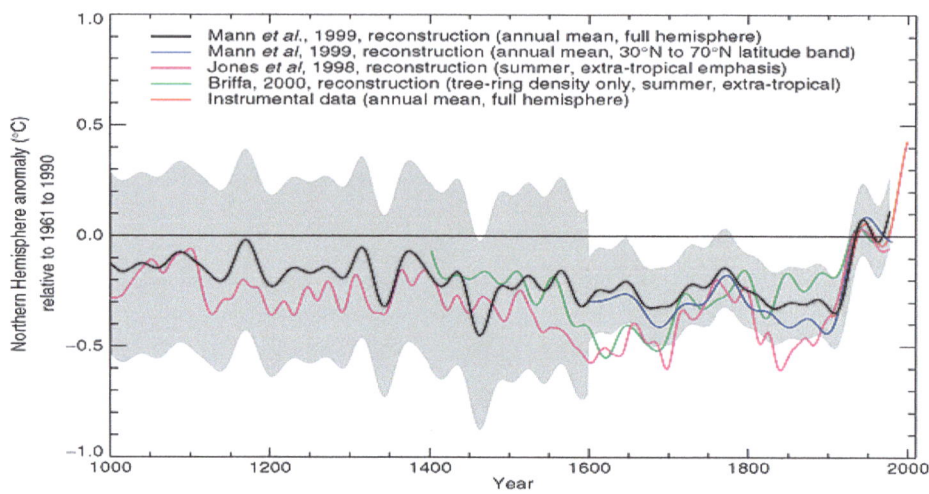

Figure 57, The "hockey Stick" Northern Hemisphere Temperature Published by Mann and his Associates (Mann et al 1999)– Primary Data used in IPCC 2001 report.

The Mann paper claimed that the 20th century was the warmest period in the last 2000 years, exceeding the Medieval Warm Period. Some geo-physicists have claimed it to be the warmest ever, which is a gross misstatement. Dr. Stephen McIntyre and Dr. Ross McKitrick performed an audit of the tree ring data used by Dr. Mann and his group to construct the Hockey Stick graph Figure 57 above and published a peer reviewed paper in Geophysical Research Letters (McIntyre and McKitrick 2005). Their conclusion was that certain data were selectively used and others selectively not used and that basic statistical methods were not used to validate their results. In particular McIntyre found that UK scientist Keith Briffa had "cherry picked" 10 tree ring data sets out of a much larger set of trees sampled in Yamai, Siberia to obtain the Hockey Stick. Further for the Medieval Warm Period the Mann group only used the low temperature data from the flawed bristlecone pine tree data. Figure 58 shows a current re-construction of the tree ring data showing no Hockey Stick large increase in Global Temperature. Congress held hearings on the controversy and two separate expert panels involving the National Academy of Science (NAS) upheld the criticisms of McIntyre and McKitrick (M & M). NAS said that the bristlecone pine data should not have been used. Emails that were intercepted from the group surfaced from the Mann group at the University of East Anglia strongly suggest their adding in of data to hide a temperature decline. Jones email stated "I've just completed Mike's Nature trick of adding in the real temps to each series for the last 20 years (i.e. from 1981 onwards) and from 1961 for Keith's to hide the decline." The Mann group have re-worked and published their new data. Mann is publically stating that his work was responsible for IPCC and Gore receiving the Nobel Peace Prize. Now Mann is suing the National Review (NR) and Mark Steyn of NR for libel and "intentional infliction of emotional distress". NR through a statement issued July 15, 2012 likened the Hockey Stick "scandal" to the Jerry Sandusky scandal and specifically stated "Except that (Sandusky) instead of molesting children, he (Mann) has molested and tortured data in the service of politicized science that could have dire economic consequences for the nation and planet." A hearing in D. C. Superior Court was held in July 2013 for a motion by National Review to dismiss the Mann suit was denied by Judge Natalia Greene. At this juncture Greene seems to side with Mann in that a research scientist's reputation is more tenuous to deformation of character than most other professions. With the NR and M & M / Mann saga still not over, one impact is certain- the public has less confidence in the man-made Global Warming scenario. Recent Duke University study refers to decade-to-decade climate change as "climate wiggles" (Brown, Li 2015).

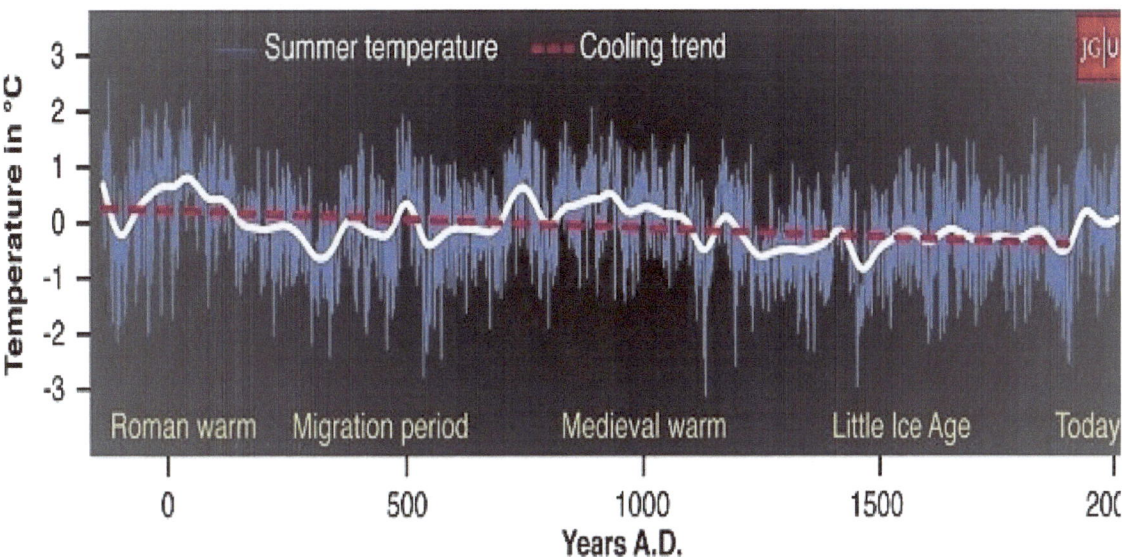

Figure 58, Estimated Global Temperature without Mann Group Data

Chapter 24 - Glacial (ice age) Region, Glacial Terminus (rise) Region and Interglacial (warm plateau) Region, Interglacial Fade Region

We first define four regions of global climate in Figure 59 based on the global atmospheric temperature and global carbon dioxide concentrations as observed in the Antarctica and Greenland ice core data.

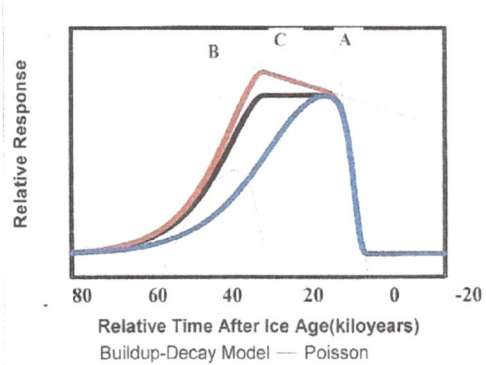

Figure 59, Regions of the Glacial-Terminus-to Interglacial-back to Glacial Cycle

With respect to Figure 59, the glacial region we define as the period when the temperature and CO_2 levels are at their minimum in the glacial-terminus-to interglacial-back to-glacial cycles. This is seen as a relatively consistent temperature below about - 5 °C for the EPICA data. The Terminus Region begins at the point in time that the temperature begins its mono-tonic increase out of the Glacial period shown in Figure 59 as region A. The Interglacial Region, we chose as the time when the temperature slope is observed to decrease into a concave slope, shown as region B. We show three shapes for region B, an upward increasing slope, a fairly constant behavior and then a negative slope with essentially no Interglacial – these three are seen in the ice core data in Figure 47. The fourth Interglacial Fade Region is the region where the temperature achieves a rapid decrease towards an obvious ice age shown as region C. The glacial periods are the minima before and after the cycle. We will find that the Interglacial Regions for both the temperature and the CO_2 exhibit positive, zero or slightly negative slopes for the cycles over the 800,000 year BP period where the 100,000 cycles occur in Figure 47.

With the wealth of past ice core CO_2 and temperature data, it is possible to directly correlate the relation between these two global climate factors. This correlation is between temperature and CO_2 values for the same time period, and sorting by CO_2 or temperature to get the variation of the other parameter with variation of the former parameter. Siegenhtaler et al [2005] in Figure 60 have used the EPICA Dome C ice core data for two periods, 0-22,000 years before present (BP) and 430,000-650,000 years BP, to plot temperature (δD - Deuterium) as a function of CO_2 concentration (ppmv) and also the Vostok ice core data from 0 – 415,000 yr BP.

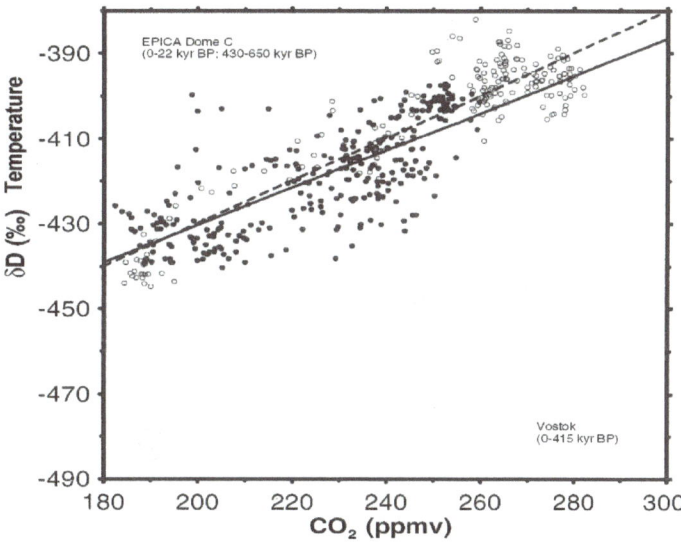

Figure 60, Siegenthaler et al (2005) Correlation of EPICA Dome C Ice Core Data for Carbon Dioxide and Surface Air Temperature

Figure 60 shows the data points and linear fits to them. The open circles and corresponding dashed line and the solid black line are linear fits to the EPICA data for the period 0 – 22,000 yr BP and 430-650 thousand years (ka) BP data, respectively. The Vostok 0-415 ka BP lower line is fit to the gray lower data. The slope for the EPICA data are 0.0837 degrees Kelvin per ppmv CO2 and 0.0946 degrees Kelvin per ppmv CO2.

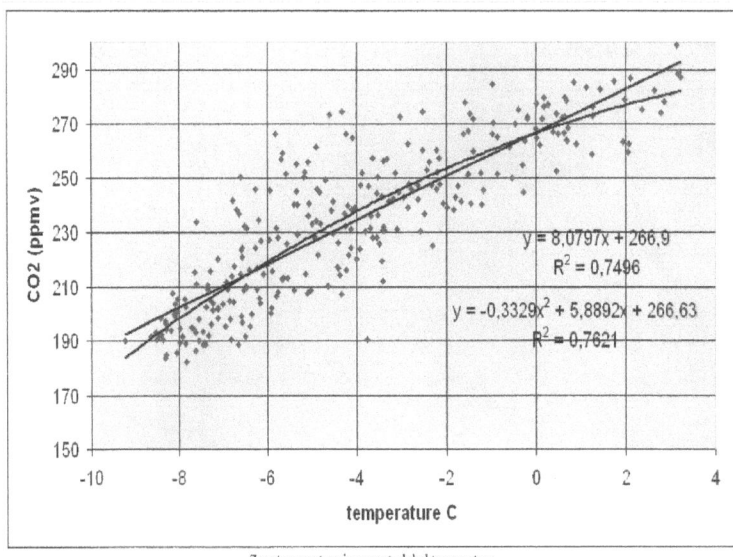

Figure 61, Petit et al 1999) Correlation of Vostok Carbon Dioxide and Surface Air Temperature.

In a similar correlation, Figure 61 provides the Vostok CO_2 vs Temperature data from 0 – 415,000 years BP published by NOAA (Petit et al, 1999) showing the linear and quadratic best fit equations. Here zero temperature is the past 50 year average. The linear fit slope is 8.0797 ppmv CO_2 per degree Kelvin temperature. This correlates to a slope of 0.124 degrees Kelvin per ppmv CO_2, a higher value than for the Siegenthaler slopes. What both these graphs show is that higher CO_2 levels correlate with higher temperature and higher temperatures correlate with higher CO_2 levels, respectively. Normally in scientific work, the independent variable is plotted as the abscissa and the dependent variable as the ordinate on the graphs. What Figure 60 shows is that Earth has experienced in the past 800,000 years a range of surface temperatures from - 9°C to +3°C (+16 °F to +37°F) and the CO_2 ranges from about 180 to 280 ppmv. But the graphs do not clarify the fundamental question of which leads and which lags and i.e. which is the independent variable (leading) and which is the dependent variable (lagging). In Figure 60, one can say that as CO_2 increases the global temperature rises i.e. CO_2 leads. In Figure 61, the CO2 ranges from about 180 to 290 ppmv and the temperature ranges from about -8.5 to +3.2 °C. One can argue that global temperature increases causes CO_2 to rise, as suggested perhaps mainly from solubility expulsion from the oceans. These graphs are also somewhat misleading in that they include all stages of Earth's climate history for those periods i.e. the glacial periods, the interglacial periods (warm periods) and the glacial terminus periods (transitions between glacial and inter-glacial periods) shown in Figure 47. Of importance is that the time scale of these three climate phases is quite different, with the glacial terminus periods experiencing very rapid changes in temperature and airborne CO_2. It is impossible to imagine the very large changes in CO_2 levels in the terminus period during pre-industrial times being from CO_2 changes from natural causes such as volcanic activity on a regular 100,000 year frequency. Use of all periods implies that the same climate forces affect all the periods. Certainly it must be readily acknowledged that the Bode type feedback factors given in equation (13) play different magnitudes in the three climate regions in Figure 59. Our concern is primarily for the behavior in the warm interglacial period.

Part VII – Direct correlations between global carbon dioxide and global surface air temperatures from ice cores and Pagani et al

In Part VII containing Chapters 26 and 27, we focus on the Antarctica and Greenland ice core data as the most consistent and reliable global climate data for the past to assess present and future global warming. We examine pre-industrial, pre-anthropogenic climate data for the present Holocene period, for past Interglacial periods and data provided by Pagani et al (2005) for the Carboniferous and Jurassic periods when life flourished at higher temperatures and higher carbon dioxide levels. We directly examine the ice core data of EPICA Dome C and compare the past Intergalactic periods with our present. We further apply the CO_2 to temperature correlation methods of Siegenthaler and Petit to the Intergalactic EPICA data only to see if CO_2 is the driving force during those periods or does extraterrestrial forces seem present and even dominate.

Chapter 26 – Ice Core Correlations Based on Our Present and Past Interglacial Plateau Regions

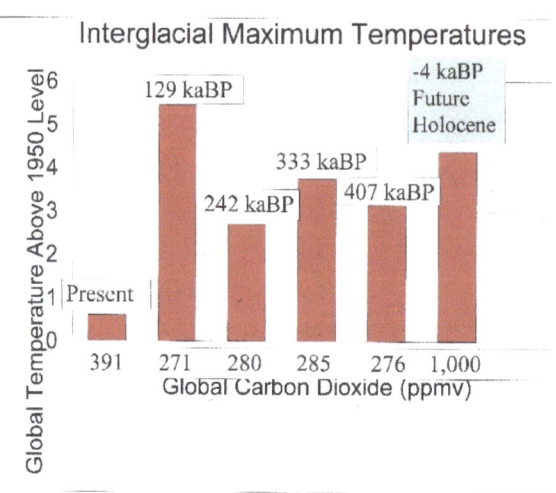

Figure 62, Maximum Global Surface Air Temperatures for Past Four and Present Interglacial Periods, EPICA data.

Numerous investigators have claimed that the present "record breaking" temperatures are the highest in the past million years – not true as seen from Figure 62 based on the NOAA published EPICA ice core data. In the figure we show the temperature above the 1950 level on the ordinate, the time data point of each atop the bar graph and the CO_2 concentration on the abscissa. This has to be then a gross violation of scientific ethics to make such claims since on one hand they use the ice core data such as Mr. Gore has done and then at the same breath make the record breaking claims. In his book, on pages 101 and 102, Mr. Gore (2006) does cite specific EPICA data and describes the flooding of the North Atlantic by cold waters from the Canadian ice sheet (waters from breakout of Lake Agassiz called the Younger Dryas Event, see Figure 77), documented in the Antarctic EPICA and Greenland GRIP ice core data. Since the temperatures in the last four most recent Interglacials have been higher than our current Holocene, we most likely can expect our Global Surface Air Temperature to reach at least 4°C above 1950 levels.

The Shapes of the Present and Past Interglacial Periods EPICA data

We shall follow the works of Siegenthaler et al (2005) shown in Figure 60 and the Petit et al (1999) NOAA work shown in Figure 61 by using CO_2 and temperature data from EPICA , but here only compare the CO_2 and temperature during the Intergalactic warm periods (IPCC is adamant that during the present Holocene Interglacial period there are no extraterrestrial forces affecting climate). Directly from the data, we find the range of global temperatures has been from about - 11°C to about +4°C above the 1950 average. Earth has recently been in cold glacial periods more than warm interglacial periods for the last 800,000 years. The overall average temperature has been - 4.58 °C with a very large variance of 11.89 °C based on the EPICA Dome C data. The maximum temperatures in the Interglacial Periods for the four most recent ice age cycles are considerably higher than the four earlier Interglacial periods.

We shall examine the ice core data in the present Holocene and earlier Interglacial periods with respect to correlation between Global Carbon Dioxide Concentrations and Global Surface Air Temperature. First, to see the past Interglacial periods, as Figures 63 and 64, we show the peak regions for the past eight Interglacial periods dating back to 800,000 years before present for both Temperature and CO_2. These data are from the EPICA ice core data and EPICA data will be used here to assess what data is relevant to correlate their interconnections. An immediate observation, that others have noted, is that only the last four Glacial-to-interglacial-back to- Glacial episodes were clearly at 100,000 year cycles. The other climate behavior from about 800,000 to 400,000 years BP appear to be a mixture of 21 and 41 year cycles in concert with the 100,000 year cycle. This will be examined in later sections.

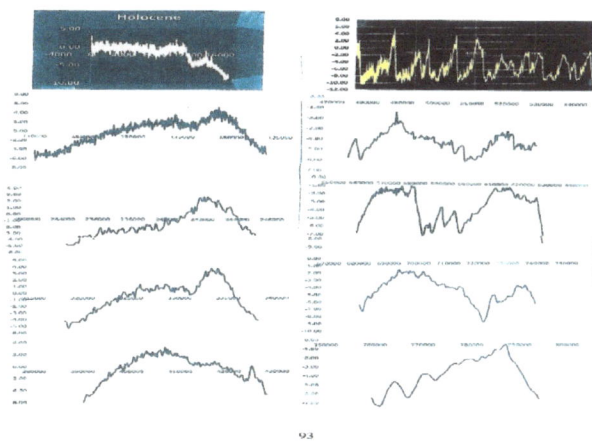

Figure 63, Peak Interglacial data for EPICA Global Surface Air Temperature

The close examination with Figures 63 and 64 show each interglacial reflecting individual characteristics. A basic approximation for the IPCC climate modeling is that the CO_2 and temperature levels during the present interglacial period (called the Holocene Period from about 13 – 0 kyrs BP), should be (and IPCC assumes to be) relatively constant except for human generated GHG forces (see Figure 2.4, IPCC Synthesis Report 2007). The Figure 2.4 assigns a Radiative Forcing factor of 0.12 W/m² to Solar irradiance out of the net of 1.6 W/m², i.e. about 7.5%. This would imply, for IPCC's analysis, that the astronomical forces variations are minimal during our present warm Holocene Period and human GHG production is the only forcing mechanism relative to our global climate.

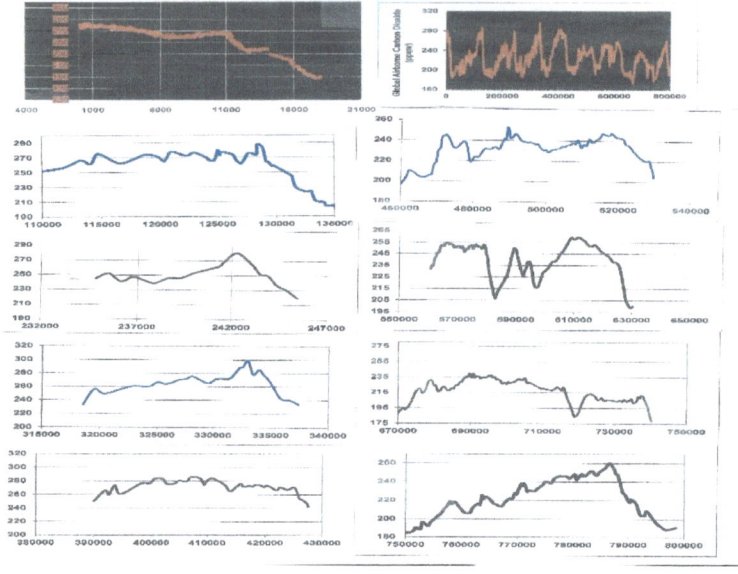

Figure 64, Peak Interglacial data for EPICA Global Surface Air Temperature

With the same logic, this premise should then be most probably applicable during the past eight warm interglacial periods shown in Figure 63. As noted above relative to the specific impact of the overall influence of the Milankovitch forces [Milankovich 1998], our contention here is that there is not yet demonstrated evidence that the Milankovitch forcing's are indeed absent during the warm interglacial phases. In Figure 2.4, IPCC assigns about 70% of the positive radiative forcing to CO_2. If then negligible astronomical climate forcing is true, then a close correlation between CO_2, with the positive feedbacks dominating from CO_2, as the driving force, should be evident. Thus a simple relation, such as Equation (15), should approximate the correlation between temperature and carbon dioxide:

$$\text{GlobalMeanSurfaceAir Temperature} \approx \alpha + \beta \times \text{Global Mean Atmospheric Carbon Dioxide} \quad (15)$$

where α is a scaling constant, β is the CO_2 to temperature correlation coefficient and $\approx$ signifies approximation. We intend here to test this first by a close examination of the past Interglacial Periods temperature and CO_2 behaviors when no human GHGs were present and their levels should be fairly constant with time.

This can be examined by comparing these past warm interglacial periods in the prior glacial-to-interglacial cycles with our present Holocene Period (Earth's present warm interglacial period) as is done in Figure 65 below, because recently our present Holocene period has seen a large increase in CO_2 from human generated GHG and the past cycles have not.

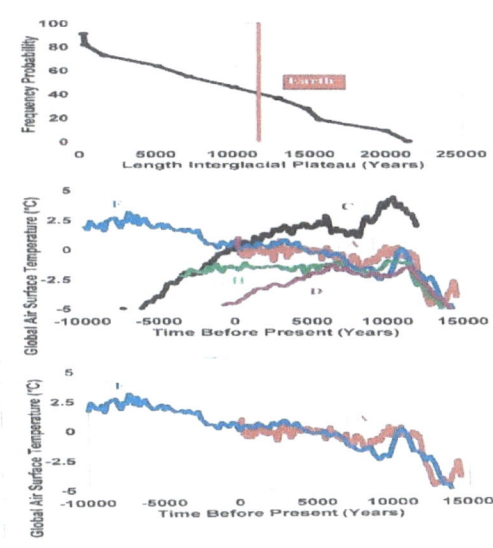

Figure 65, Examination of the Interglacial Periods for our Present Period and Others that Resembles Our Present.

The top panel of Figure 65 examines the estimated frequency distribution of the length of the warm interglacial periods (using the criteria in Figure 59 to estimate the duration for the Interglacial Period) for the past 11 glacial-interglacial cycles. The red line denotes the present length of our, yet incomplete, interglacial cycle (Holocene). We see that some past interglacial periods were much longer and some much shorter with a very large range of values. In the middle panel of Figure 65, we have examined the shapes of some of the Air Surface Temperatures in the interglacial periods most similar to our present. Although, as noted, the shapes appear quite similar in Figure 47 on a large scale (as Mr. Gore has implied), in Figures 51, 52 and 53 showing the work by others, they have shown from close examination that they are very different. This is indicative of the very large variations in interglacial lengths in the top panel of Figure 65. As noted, the middle panel of Figure 62 presents the specially selected interglacial periods that are most similar to our present Holocene period, identified by their approximate dates that the terminus periods occurred i.e. labeled A (Holocene – 13 kyrs BP), C (132 kyrs BP), D (218 kyrs BP), F (430 kyrs BP) and H (580 kyrs BP). The time scales are matched at the terminus period slopes for synchronization of the interglacial time scales. We used Figure 64 Interglacial graphs to select these Interglacial periods. What we are looking for is any prior interglacial that had a similar Global Climate to our present Holocene. The interglacial period remarkably similar to the present is the one labeled F (430 kyrs BP) as seen by close examination in the bottom panel of Figure 65 which had a very long Interglacial Period. We do not know how long our present warm interglacial period will last {Loutre and Berger [2000] suggests a long one} but it is also conceded by many geo-scientists that it is highly probable that in the future we will have another ice age driven by astronomical effects. Does this mean, due to the similarity to the 430 kyrs BP interglacial, that our warm period will be as long? This has been examined by others. We will examine such a premise in later sections.

In Chapters 27 and 28, we shall examine the correlation between CO_2 concentration and surface air temperature for the warm Interglacial Regions. First we examine climate behavior for only the pre-industrialized present Holocene interglacial period and then for past warm interglacial periods. This thus tests the IPCC premise that the CO_2 and temperature correlations are consistently quasi-

85

linear and/or Bode feedback equation compatible,5 given by their equation (14) above, when applied to these past warm interglacial periods and our present pre-industrial cycle when, in these cases, human GHG forcing were not present.

For Global Climate purposes and GHG effects, we can consider pre-industrial period as before world production of carbon dioxide was accelerated. Figure 66 provides world CO_2 production rates dating back to 1860 AD.

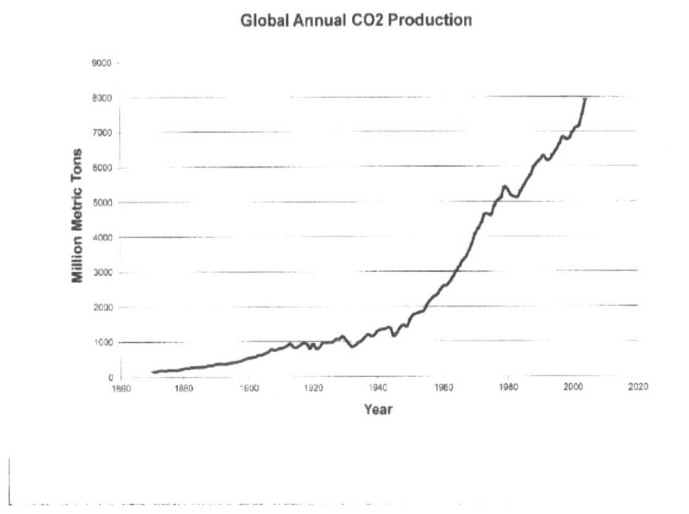

Figure 66, Global Annual CO2 Production

In Figure 66, we see that CO_2 production in 1870 at the beginning of the graph, was minimal. For our analysis, we use the EPICA ice core data for both CO_2 and temperature. We correlate for each ice core time data point the corresponding CO_2 and temperature values, but only during the warm interglacial periods whereas, in Figures 60 and 61, their correlations covered all phases of the glacial-to-interglacial-back to-glacial periods. It is possible graphically to scale either temperature or CO_2 levels to fit the glacial terminus regions to the corresponding levels of the other parameter.

To examine the effect of one upon the other, we here do this for both temperature as a function of increasing CO_2 and CO_2 as a function of increasing temperature. First in Figure 67 we do this for the recent Holocene Period (our present interglacial warm period). In the subsequent sections and Figures 68, 69, 70 and 71, we will examine Interglacial regions for periods respectively from 130 to 110 kyrs BP, 430 to 410 kyrs BP, 575 to 565 kyrs BP and 627 to 610 kyrs BP. Siegenthaler et al and Petit et al found linear regressions gave good fits to their data. To obtain the best linear fit to the data, we have applied the Method of Maximum Likelihood Estimation [Savage and Papworth, 2000] which has been used in many research studies and by the author on previous research (Leonard 2012).

As the time scale for the CO_2 data was not exactly the same as that for the Temperature data, interpolation on the dataset was undertaken to obtain both at equal time intervals. Further, the number of line items in the CO_2 file was only 1057 compared with 5788 line items in the Temperature file. The interpolation was performed stepwise as follows:

1. The age assignment of the Temperature file was retained.
2. The file was scanned by age from the Temperature file choosing AgeT01.
3. The CO_2 file was scanned to find the age nearest to Age01 but not greater, choosing AgeC01
4. The next age in the CO_2 file was selected and identified as AgeC02
5. Values of CO_2 for AgeC01 and AgeC02 were read from the CO_2 file and identified as C1 and C2 respectively.
6. The value of CO_2 assigned to the resulting file was calculated.

$$\text{Value} = (\text{AgeT01-AgeCO1})/(\text{AgeC02-AgeCO1})*(C2-C1) +C1$$

7. The value was entered into the file named GWLIN.TXT opposite AgeT01.

The new file having the original values for age and temperature had a third column with this Value inserted.

Calculating on the values of the temperature and CO_2 was then greatly facilitated by this interpolation.

A test was conducted using the 2 values of CO_2 before and after the selected age and calculating finding the new value at the selected age using a second order function. This was observed to make the file noisier than did the linear interpolation. That file has not been used in further calculations.

Holocene Period Correlations

Holocene and Past Period Correlations

We have chosen 0.20 kyrs BP as the latest period for the Holocene Period analysis since NOAA data seen in Figure 66 shows that at that time man-made CO_2 concentrations, globally were minimum. Our analysis presented in Figure 67 of pre-industrial Holocene Period is for the periods 2.0 to 0.028 kyrs BP and for 5.0 to 2.0 kyrs BP. Panel A shows the CO_2 and scaled Surface Air Temperature to fit the CO_2 curve. Panel B provides the expanded curves from Panel A for Atmospheric CO_2 as the top red curve, a smoothed mono-tonic fit in black, the scaled Surface Air Temperature in the fine blue curve and the bottom red curve as the smoothed temperature. In Panel B, the CO_2 and temperature show an excellent fit between 5800 and 6800 years BP and then considerable divergence at later times approaching present with the CO_2 level increasing and the temperature remaining relatively constant. Panel C provides the corresponding Air Temperature data point values for values of CO_2 similar to Figure 60 for all EPICA data. IPCC has estimated that at least 70% of the Global Warming effect is from carbon dioxide with the other 30% from other gases such as methane and other feedback factors. We therefore attempt to obtain a reasonably good accuracy with the premise that CO_2 causes Global Warming by a linear relation between CO_2 and Global Surface Air Temperature approximated by Equation (15). The black arrow shows the CO_2 value at 2.0 kyrs BP. The linear best fit to the data are

For 2.0 – 0.038 kyrs BP   Temperature = 1.235 – 0.0046 x CO2 Concentration
For 5.0 – 2.0 kyrs BP   Temperature = 0.04989 – 0.00042 x CO2 Concentration

In Panel D, we show the Surface Air Temperature and the scaled CO_2 equivalent variation to fit the temperature data. From this, as Panel E we obtain a plot of CO_2 data point values for corresponding Surface Air Temperature values similar to Figure 61 for the Vostok data. The linear best fit to the data points are

87

For 2.0 – 0.038 kyrs BP   CO2 Concentration = 287.70 – 2.326 x Temperature

For 2.0 – 5.0 kyrs BP   CO2 Concentration = 351.07 – 0.1372 x Temperature

We see that the correlations for both the period 5.0 to 2.0 kyrs BP have negative slopes because the CO_2 levels are increasing faster than the corresponding temperatures. This again suggests that pre-industrial CO_2 level increases have not driven global temperature increases.

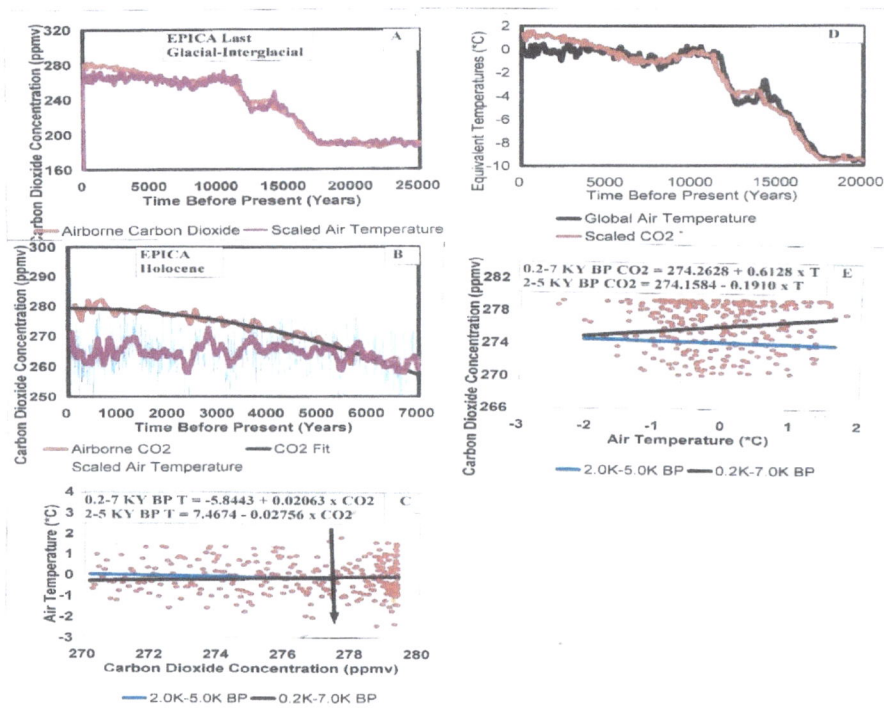

Figure 67, Holocene Period Correlations Between EPICA data Carbon Dioxide and Surface Air Temperature.

Past Interglacial Periods Correlations

Presented as Figures 68, 69, 70 and 71 are the similar analyses for the interglacial periods respectively from 130 to 110 kyrs BP, 430 to 410 kyrs BP, 575 to 565 kyrs BP and 627 to 610 kyrs BP, respectively. In each of the Panel A's and C's, the Carbon dioxide concentrations are in solid red and the black curves are the temperature. The solid black linear lines in the Panel A's show the range of EPICA data points used in the correlations (heavy red), the smoothed best fit (black line) and the scaled Surface Air Temperature (light black line). In each Panel B, the direct correlation is provided between observed CO_2 levels and Air Temperature data values. Panel C provides the Surface Air Temperatures and scaled equivalent CO_2 concentrations as was done for the Holocene in Figure 67, Panel D. Then as Panel D here for each, the CO_2 concentration data point values are plotted for each Surface Air Temperature value similar to Panel E of Figure 67.

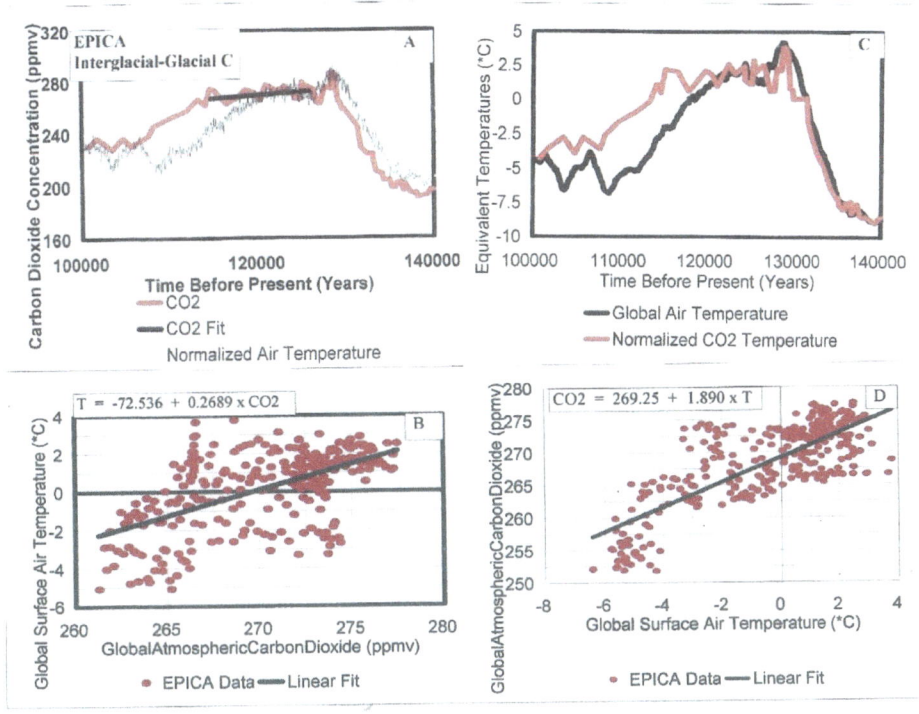

Figure 68, Correlations Between EPICA Data Global Carbon Dioxide Concentrations and Global Surface Air Temperature for 130 ka BP Interglacial Period

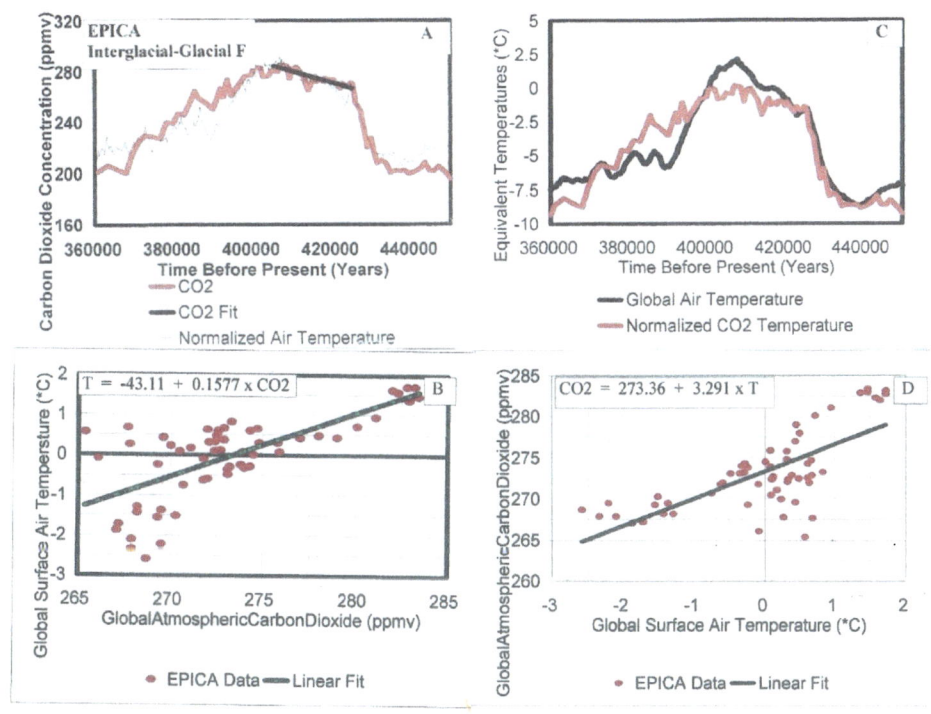

Figure 69, Correlations Between EPICA Data Global Carbon Dioxide Concentrations and Global Surface Air Temperature for 430 ka BP Interglacial Period

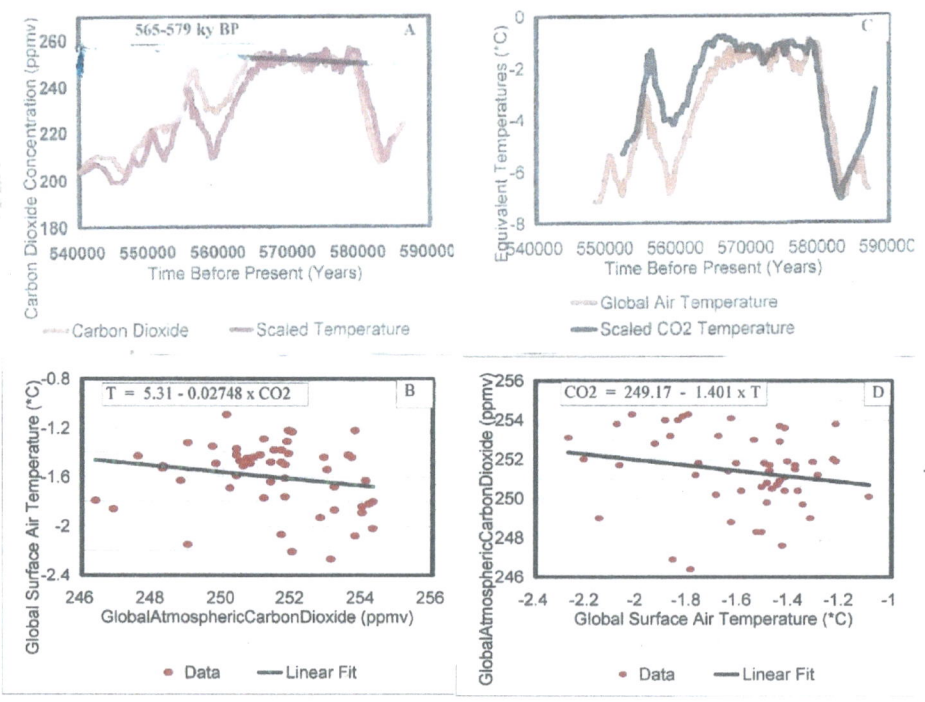

Figure 70, Correlations Between EPICA Data Global Carbon Dioxide Concentrations and Global Surface Air Temperature for 570 ka BP Interglacial Period

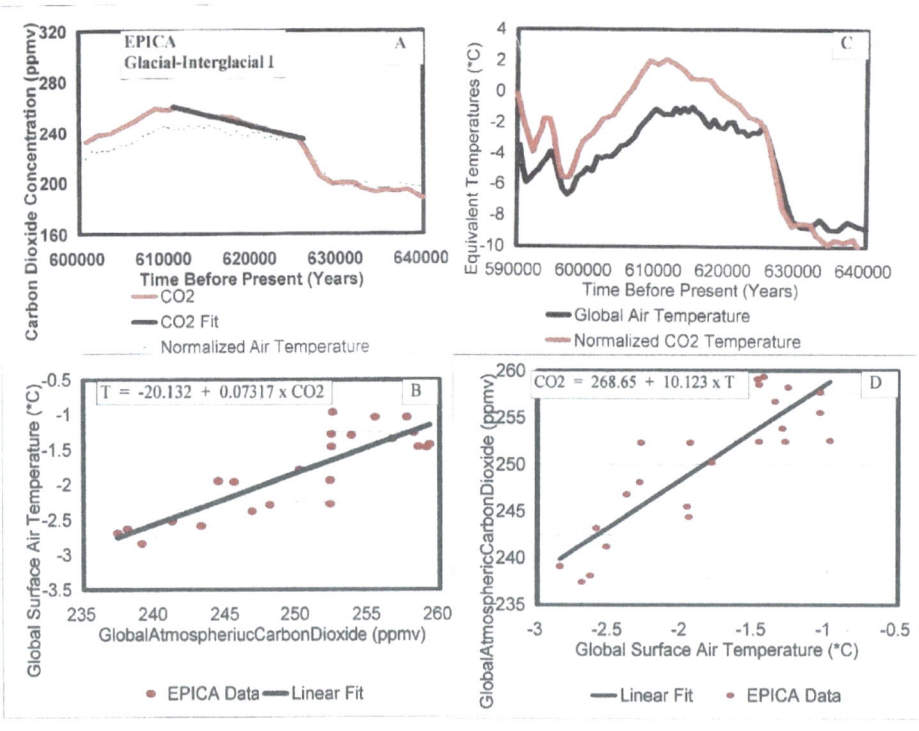

Figure 71, Correlations Between EPICA Data Global Carbon Dioxide Concentrations and Global Surface Air Temperature for 630 ka BP Interglacial Period

90

Finally, in Figure 72, we have compared the Global CO_2 and Global Temperature for the Holocene period from 1958 to 2003 and obtained a linear fit to the bottom left graph of Temperature versus CO_2.

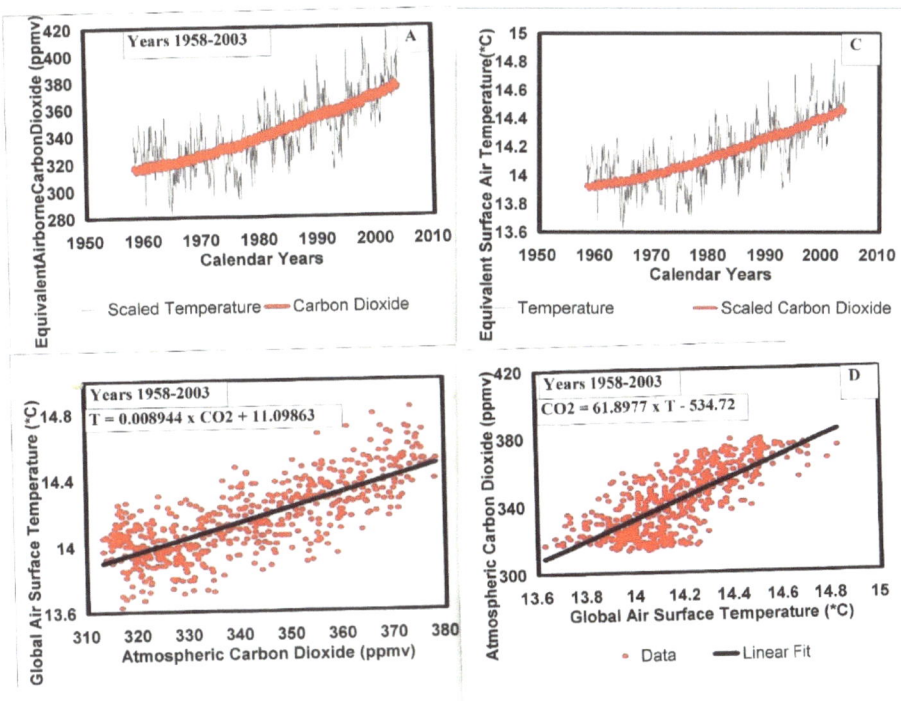

Figure 72, Examination of three Holocene Time Periods of Global Carbon Dioxide and Global Temperature NOAA National Climate Data Center Data

The best fit linear equations, as shown in the graphs, are

For 130 to 110 kyrs BP	Temperature = -72.53 + 0.2689 x CO2 Concentration
For 430 to 410 kyrs BP	Temperature = -43.11 + 0.1577 x CO2 Concentration
For 575 to 565 kyrs BP	Temperature = 5.312 - 0.02748 x CO2 Concentration
Lag of CO_2 by 700 yrs	Temperature = 3.660 - 0.02077 x CO2 Concentration
For 627 to 610 kyrs BP	Temperature = -20.13 + 0.07318 x CO2 Concentration
For years 1958-2003	Temperature = 11.09863 + 0.008944 x CO2 Concentration

The best fit relations for CO_2 Concentration as a function of increasing Air Temperature are

For 130 to 110 kyrs BP	CO2 Concentration = 269.25 + 1.890 x Temperature
For 430 to 410 kyrs BP	CO2 Concentration = 273.36 + 3.291 x Temperature
For 575 to 565 kyrs BP	CO2 Concentration = 249.17 - 1.401 x Temperature
For 627 to 610 kyrs BP	CO2 Concentration = 228.65 + 10.123 x Temperature
For years 1958-2003	CO2 Concentration = -534.72 + 61.8977 x Temperature

What is very obvious from these data analyses of present and past interglacial periods, when human GHGs were not generated, is that there is a very large variation shown in the best linear fits. If indeed CO_2 is dominant, an approximation to IPCC's expressed primarily dependence on CO_2 as the major GHG would be a single linear relationship given by Equation (15) above. Further, we

91

show for each warm interglacial period that there is a very large variance to mean of the data about the linear best fit values. Figure 73 provides a summary of the above provided best linear fit curves by the use of equation (15) showing an indisputable lack of correlation between carbon dioxide concentration variation and global temperature.

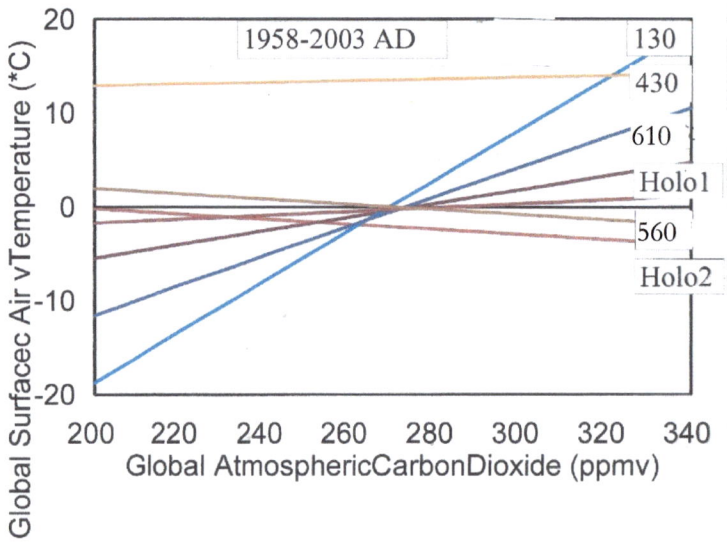

Figure 73, Best Fit Correlations Between Carbon Dioxide Concentration and Surface Air Temperature Using EPICA Ice Core for Four Interglacial Warm Periods, Our Holocene Interglacial Periods 2.0 to 0.038 ka BP (Holo1) and 5.0 to 2.0 ka BP (Holo 2) and Meteorological Data (Global data from NOAA National Climate Data Center) For 1958 to 2003 AD.

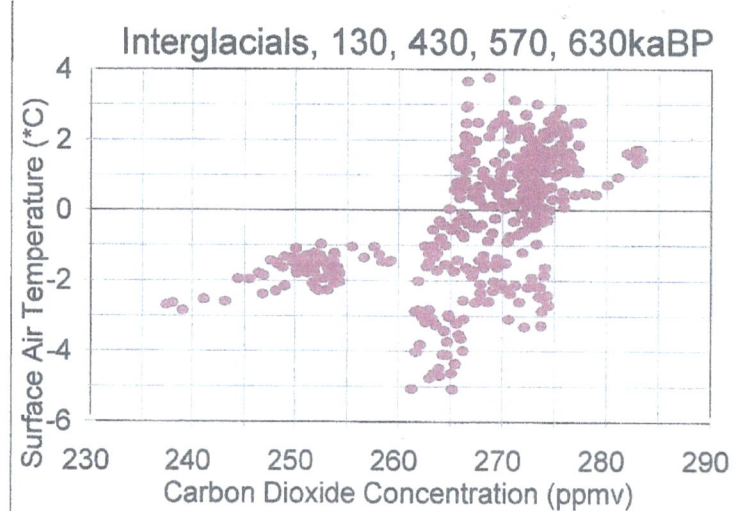

Figure 74, The Temperature Distribution Data for all Four Interglacial Periods Showing Failure of Linear Correlation Between Global Temperature and Global Carbon Dioxide as Would be Expected from Approximation Equation (15) and IPCC

The Figures 73 and 74 clearly validate that, for the past warm Interglacial periods, carbon dioxide forcing has not driven global temperature variation and has not dominated Global Climate.

The EPICA data has been normalized to the average Global Temperature for the past 1000 years. The 1958 to 2003 data, in Figure 72, are for current Global Surface Air Temperatures. The data analysis supports a premise that during the warm interglacial periods there is minimal correlation between the change in global temperature and the change in atmospheric carbon dioxide concentrations. Also the directly recorded, real time present data (1958-2003) in Figure 72 shows an extremely small slope of 0.008944 degrees temperature increase per carbon dioxide ppmv increase compared to the slopes in the past interglacials. If Global Atmospheric Carbon Dioxide Concentration is doubled from 390 ppmv to 780 ppmv, the 1958 to 2003 AD data would suggest a temperature increase of 3.49 °C which is in the IPCC range in Equation (15). However, we cannot ignore the other past Interglacial data which suggests, in their order, that the doubling of Global CO_2 would produce a very wide range of temperature changes i.e. 493.2 °C, 92.5 °C, -57.4 °C and 19.0 °C Global Temperature changes for the 130, 430, 575 and 627 ka BP cycles, respectively. For the Figure 67 Holocene data we would have -10.7 °C and +8.0 °C for the 2 - 5 ka BP and 0.2 - 7 ka BP Holocene periods, respectively for CO_2 doubling. Thus, for this Holocene interglacial period, there is minimal CO_2 influence on global temperature, where large CO_2 changes have occurred, as compared to pre-industrial periods in the other figures. We therefor again verify that CO_2 does not dominate Global Temperature during the Interglacial warm periods as well as the Glacial and Terminus periods and that extraterrestrial astronomical forces (technology not yet able to quantitatively predict to good accuracy) dominate Global Climate even during the Interglacial warm periods.

Change in CO_2 and/or Temperature From Prior Change in Other

A less direct but still significant correlation exists between CO_2 and/or temperature for subsequent changes in CO_2 and/or temperature as a function of prior change in the other. Here we examine subsequent change in one parameter (CO_2 or Temperature) 1, 2, 4 and 5 years after a given value of the other (Temperature or CO_2) is observed. Figure 75 provides the relative change data and the best fit linear equations. The equations are

Temperature Change 1 Year = -0.18058 + 0.000564 x CO2 Concentration

Temperature Change 2 Years = -0.2166 + 0.000672 x CO2 Concentration

Temperature Change 4 Years = -0.1830 + 0.000572 x CO2 Concentration

Temperature Change 5 Years = -0.1662 + 0.000522 x CO2 Concentration

CO2 Concentration Change 1 Year = -16.5094 + 1.2625 x Temperature

CO2 Concentration Change 2 Years = -8.8417 + 0.7203 x Temperature

CO2 Concentration Change 4 Years = -8.6842 + 0.7094 x Temperature

CO2 Concentration Change 5 Years = -8.4303 + 0.6919 x Temperature

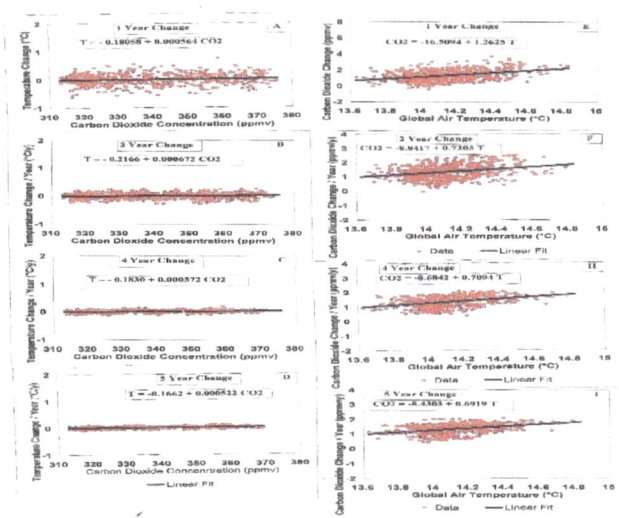

Figure 75, A measure of the Effect of Increase in CO2 or Temperature on the subsequent increase in the other parameter.

What is observed, with the Figure 75 data, is that increasing global temperature causes a significant increase in CO_2 concentrations in the next subsequent years, however increase in CO_2 levels causes minimal increase in global temperature in the next subsequent years. This suggests that CO_2 increases lags and responds to temperature increases.

Chapter 27 – Other Data – Analysis of Paleogene High CO2 and Temperature Data (Pagani et al 2005)

Our pre-industrial CO_2 levels had largely stabilized since about 9 million years BP to a range between 180 and 290 ppmv even during Interglacial cycles. We know that atmospheric CO_2 can cause a very large global greenhouse temperature effect from our observations of surface temperatures on our neighbor planet, Venus, and what we estimated as CO_2 levels on early Earth in Figures 33 and 35. In Figure 33, Panel D, we have estimated global temperatures back to 2.4 billion years BP based on the estimates of Lowe and Tice (2004). Probably the best combined early data of temperature and CO_2 are the NOAA data of Pagani et al (2005) where they obtained temperature and CO_2 data from about 45 to 5 million years BP from deep sea cores. This period was after the dinosaur extinction event (DEE) and then the Paleocene-Eocene Thermal Maximum (PETM) excursion, which is not understood, showing very high CO_2 levels in very short time periods (Pagani et al 2006). One explanation for PETM is a period of extremely active volcanism. Since PETM occurred immediately after DEE, the asteroid impact for DEE could have shaken the Earth's mantle to provoke the volcanism. We show the Pagini et al (2005) CO_2 data as a function of time in Figure 33, Panel B and combined with the Berner and Kothavaka (2001) data in Figure 33, Panel C. As Figure 76, we examine their data in terms of Atmospheric Temperature data as a function of higher CO_2 concentration data up to 1,800 ppmv where the GHG effect should be evident, by numerically sorting the CO_2 levels to obtain this correlation i.e. the variation of Global Temperature with before present industrial caused higher levels of CO_2.

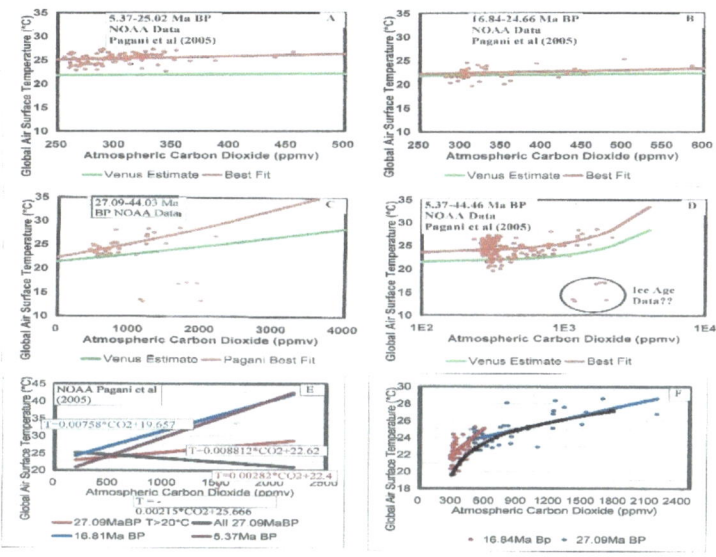

Figure 76, Million Year BP Data of CO2 and Air Temperature from Pagani et al (2005)

We show the three different sets of NOAA data for the different time ranges in Panels A, B and C. and the fit to all the data with the abscissa in log scale in Panel D. The green solid line is an estimate based on the Venus greenhouse behavior. This is first using the NASA Venus average surface temperature of 480 °C and assuming an inverse square (distances from the sun) decrease in solar luminosity at the top of the atmospheres to obtain an equivalent Earth temperature of 253 °C with an 98% CO_2 atmosphere containing 1.996 x 10^{20} kg of CO_2 (assuming all the carbon given in

Table 2 were initially as CO_2 in the atmosphere from volcanic emissions). We assume a green-house exponential attenuation of the reflective radiation beginning with the 98% CO_2 and the present day 380 ppmv of CO_2 and Global Earth temperature of 21.4 °C.. This yielded an attenuation coefficient of 1.7×10^{-5} per ppmv CO_2. For the three Pagani et al (2005) data sets, we obtained for zero CO_2, temperatures and attenuation coefficients of 24.2 °C and 1.45×10^{-4}; 21.5 °C and 1.35×10^{-4} and 22.3 °C and 1.23×10^{-4} respectively for the Panels A, B and C best fits. For the best fit of all the data in Panel D, we obtained 23.4 °C and 9×10^{-5} °C per ppmv. The actual best fits shown in red are

9 5.37 – 25.02 Ma BP   Temperature = 22.636 + 0.00879 x CO2 Concentration

16.84 – 24.66 Ma BP   Temperature = 19.672 + 0.00756 x CO2 Concentration

27.09 – 44.46 Ma BP   Temperature = 22.456 + 0.00279 x CO2 Concentration

All 5.37 – 44.56 Ma BP Temperature = 23.842 + 0.00148 x CO2 Concentration

In the 27.09 Ma BP data analysis, we considered the six data sets with temperatures below 20°C as outliers which could be from cooling events such as large volcanic eruptions or an asteroid impact. We did perform a best fit for that case in Figure 76E, where we show the fits for all cases. Figure 76E suggests two observations, that the 5.37 Ma BP data should be more in agreement with the other two sets at the time period near 25 Ma BP, where we see its best fit value at 25 Ma BP is about 3°C higher than the other two. In fact, the 16.84 and the 27.09 Ma BP sets are in such good agreement, we graphed as Figure 76F the two together. What we observe is a Bode type negative feedback similar to that proposed by IPCC. Here this would be a negative feedback with increasing CO_2 (and temperature). The red and blue straight lines are the best fit linear curves for each and the solid black curved line is the best quadratic fit.

Note that the CO_2 to temperature coefficients of 0.00756 and 0.00879 are comparable to the low value of 0.008944 for the 1958 to 2003 data in Figure 72. In Panel D, we show the Intergovernmental Panel on Climate Change (IPCC 2007) estimate of CO_2 effect on global surface air temperature using their CO_2 doubling range of 2 to 4.5 °C per doubling of CO_2. We also show the Pagani et al and our attenuation theory curves. We thus estimate a 0.345 °C increase from doubling of CO_2 from these data. Using the Panel D fit for all the Pagani et al data, an increase in Earth CO_2 concentration by a factor of 5 (from 360 to 1800 ppmv = 1440 ppmv change) would incr9ease Earths average Global Temperature by 1.01 °C. We have fit an exponential function to Pagani et al data but within the range of temperatures and CO_2 variations that we expect in the next few centuries the variation can be linearly approximated giving a temperature rise rate of 0.09 °C per 100 ppmv CO_2 increase for small changes. We note that in examining the Rise Region of the Interglacial cycle that an increase of 1 °C in temperature resulted in an increase of 7.022 ppmv in CO_2 level, thus an extraterrestrial forced 0.142 °C rise in temperature results in a 1 ppmv CO_2 change with temperature as the driving force for change during the Terminus Rise Region. But above we show that the positive feedback of increased Global Warming for increased airborne carbon dioxide is very small. In the previous section, we premised a 2.5 to 3.5 °C temperature rise during the remainder of our present Interglacial cycle. If, from human activities, the CO_2 level went to 1800 ppmv, we would see at most an additional 1 °C. We conclude that Global Surface Air Temperature is very insensitive to CO_2 change and does not affect the Interglacial transition behavior as was also shown in Figure 73. Since CO_2 appears to lag temperature, other extraterrestrial forces must drive the interglacial cycles dominating Earth's climate on a long-term scale.

PART VIII – THE YOUNGER DRYAS (YD) AND 8.2 ka BP COLD EVENTS

In Part VIII, which includes Chapters 28, we examine ice core data for disruptions in Interglacial Plateau periods by variation of the thermohaline Conveyor Belt North Atlantic ocean circulation. We examine the Younger Dryas Event and the 8.2 ka BP Event.

Chapter 28 – Examination of Sea Level, Sea Surface Temperature and Ice Core Data For the Younger Dryas (YD) and 8.2 Kilo-year Events

During the Terminus Period leading to our present Holocene Period a sudden resumption of ice age type climate was observed, called the Younger Dryas Event. For the Greenland GRIP Ice Core Data, the Surface Air Temperature data has been separately analyzed for the Younger Dryas Event (data from 14,016 to 10,942 years BP) and the 8.2 ka Event (data from 8,517 to 8,003 years BP) and has been published by NOAA. It is suspected that both of these abrupt, cold spell events that were experienced world-wide were caused by a sudden burst of cold fresh water into the North Atlantic causing a disruption of the Atlantic Ocean North-South thermohaline circulation. We will discuss the thermohaline circulation and its effect on the North Atlantic Deep Water formation and the world-wide Great Ocean Thermal Conveyor Belt in a separate section below. We here wish to examine these data with respect to temperature effects on other world proxies such as Sea Surface Temperature and Sea Level. Figure 77, Panel A provides the NOAA data for GRIP (in green), the SST (in red) and the Sea Level (in blue) for the YD and 8.2 events.

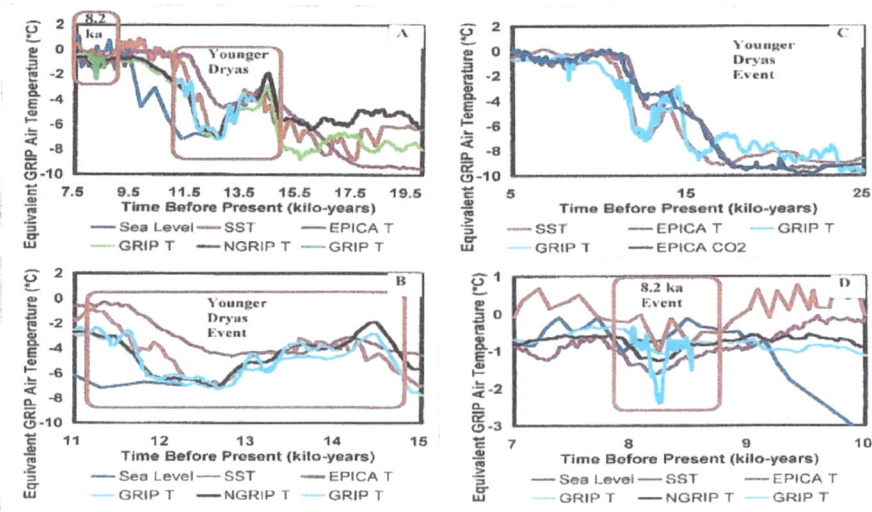

Figure 77, Analysis of Younger 39Dryas and 8.2 ka Cold Events and Correlation With North Atlantic and Lake Agassiz Cold Water Surge

Also, shown are the Surface Air Temperatures (SAT) as measured by the EPICA ice core (EPICA T in purple) and the North GRIP ice core (NGRIP T in black). In Panel B, we examine the YD Event. We see that the SST sudden decrease coincides with the sudden decrease in both the Surface Air Temperatures for Greenland GRIP T and NGRIP T but the Antarctica EPICA T is effected much less. The Vostok ice core data show a moderate affect also which mimics (not shown) the EPICA data. This suggests that the event was a SST event and primarily localized to the Northern Hemisphere, affecting the Surface Air Temperature at the Greenland GRIP T and NGRIP T sites (the SST would lag temperature if it were not a cold water flooding event, which it does not). As Panel C, we show the EPICA carbon dioxide data for the YD period, showing the CO_2 concentration following behind the EPICA T, as we studied above, but affected to a lesser extent even than the EPICA T, thus the YD Event could not have been caused by CO_2 effects. Although the source of the event was in the Northern Hemisphere, it did affect the climate world-wide. The stalagmite Hula Cave, China data reflected a slight temperature reversal at that time. So the premise that the YD Event was from the break-out of Lake Agassiz from the Laurentide Ice Sheet appears to be a valid premise although there are other less probable theories such a massive volcano eruption or a giant impact event (but temperature would lag CO_2 for both which it did not).

Figure 78, Four Phases of Melt of Laurentide Ice Sheet and Breakout of Lake Agassiz Cold Water Surge

The lower right panel of Figure 78 shows that at the maximum Glacial Period, Canada was buried under thousands of meters of ice all the way past the Great Lakes and to New York City where Long Island was created as a glacial moraine. The geology provides for the ground elevation to slope North from the Great Lakes towards the Hudson Bay (This is because the Earth's crust was depressed due to the weight of the ice sheet.). The upper right panel shows then the ice sheet melt in the Southern portion, the drainage Northward and with the unmelted Northern, colder portion as an ice dam. In fact Hudson Bay was formed as a depression from the weight of the Laurentide Ice Sheet that apparently formed during each ice age. The upper left panel shows the estimated details of the lakes breakout with the three red arrows. As Figure 79, we show the lake at its fullest extent.

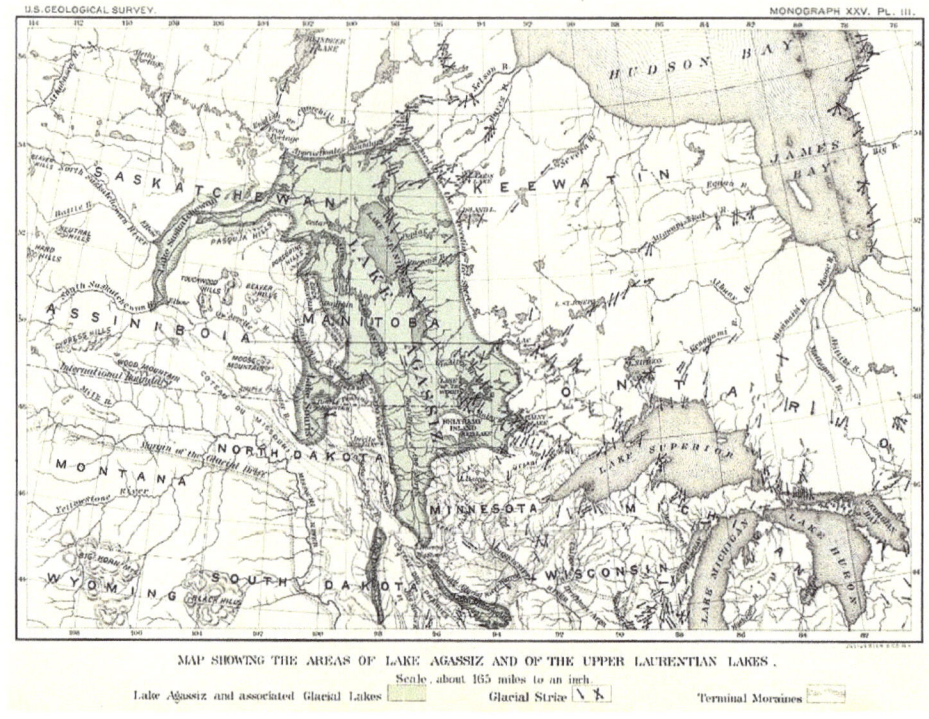

MAP SHOWING THE AREAS OF LAKE AGASSIZ AND OF THE UPPER LAURENTIAN LAKES.
Scale, about 165 miles to an inch.

Lake Agassiz and associated Glacial Lakes   Glacial Striae   Terminal Moraines

Figure 79, Map Showing Areas Around and Including Ice Age Lake Agassiz

In Panel D of Figure 77, we show the 8.2 ka Event proxy data. The data shows that the event must have been rather localized to the GRIP site since the NGRIP T responded much less and similar to EPICA T and Vostok response. The 8.2 ka Event was much smaller in temperature decline and duration than the YD Event but more abrupt – note how the YD Event initial cooling was gradual in transition over a period of about one thousand years whereas the 8.2 ka transition occurred over only about 100 years. It may be that for the YD Event, Lake Agassiz was a more gradual leakage around or under the Laurentide Ice Sheet that more slowly emptied it and the 8.2 ka was an abrupt, but smaller break-out. It is known that the ice volume increased during the YD Event lower temperatures because the North American and European human cultures (archeologist records), that had moved North in the warming period from 18 ka BP to 14.5 ka BP, retreated Southward again during the YD cooling period. Then the 8.2 ka lake, given the name Glacier Lake Ojibway, would be smaller. It is possible, but unlikely that the 8.2 ka Event could have been caused by the break-out of the Baltic Sea from the last ice age European Ice Sheet thawing. We will discuss the YD and 8.2 ka Events below.

Are There Other Cold Water Surge Events Like YD and 8.2 ka Events in the Past Interglacial Cycles?

The Arctic Ocean has been nearly enclosed now by the Northern Hemisphere land masses on Northern Canada, Alaska, Greenland, Scandinavia and Russia for over 10 million years with 20 and 40 ka ice age cycles and later to the present 100,000 year cycles. Because the apparently cold water surges of Younger Dryas and 8.2 ka Events, from Lake Agassi and Lake Ojibway and other possible terminus region melts, may impact on our future climate if they occur again, it is important to examine in the past whether these type cold water surges have been frequent during the Terminus Rise Regions and early Interglacial Plateau Regions in other ice age cycles. In Figure 65, we note that

100

the 130 ka and the 330 ka Interglacial plateaus show a terminus rise to maximum high temperature peaks and then sharp declines before any Interglacial plateau is achieved. That is instead of smooth transitions of the Terminus Rise Regions into the early Interglacial Plateau Region, the transitions show a significant brief negative decline of about 3 ° C in about 1,000 to 2,000 years, just as did occur just before the 8.2 ka Event at the beginning of our Interglacial Period in our present Holocene Plateau. In further analysis here, we examine the Surface Air Temperature data of EPICA in Figure 62 and compare with the Sea Surface Temperature of NOAA in Figure 49.

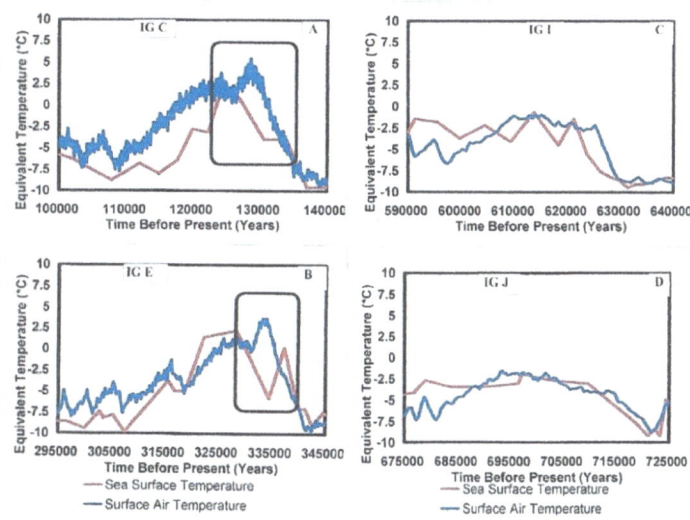

Figure 80, Comparison of NOAA Sea Surface Temperature with EPICA Surface Air Temperature for Evidence of Cold Water Surge Creating Abnormal Interglacial Plateau Behavior

What we observe for both 130 ka and 330 ka Terminus is that each appears to have experienced a sudden Sea Surface Temperature decrease, shown in red in Figure 80, as an abrupt change in North Atlantic Sea Surface Temperature briefly preceding the sharp decrease in Surface Air Temperature. In Panels C and D for both these other two Terminus at 630 ka and 720 ka, we do not see the same behavior for the Sea Surface Temperature nor do we observe the abrupt decrease in Surface Air Temperature at the beginning of the Plateaus, instead we see the expected lag as would occur if the SST was driven by the Surface Air Temperature. The correlation in Figures 80, Panels A and B is we expect to be apparently from cold water surges in the North Atlantic, also most probably from a break-out of trapped glacial lake water either from the prior Laurentide Ice Sheet or an European Ice Sheet from the waning Ice Age glaciation. Note that both in Panels A and B that the Interglacials achieved very high Surface Air Temperatures (3.5 to 5.0 °C peak) whereas the Panel C and D Interglacials were much cooler (-2.0 to 0 °C peak), thus perhaps not warm enough to completely melt the ice sheets and release the cold glacial waters. At these high temperatures there is no reason not to expect a similar Lake Agassi breakout

PART IX – PLATE TECTONICS CREATION OF INTERGLACIAL AND ICE AGE CONDITIONS

Part IX contains Chapters 29, 30 and Discussion which presents how the Ice Ages developed from the closure of the Pacific to Atlantic Oceans Atrato Seaway in the Panama Isthmus and the positioning of Antarctica at the South Pole, hence the establishment of the Thermahline North / South currents transporting warm water from the Caribbean and Gulf of Mexico and how disruption causes the Ice Ages.

Chapter 29 – Plate Tectonics Creation of Gulf Stream and Thermohaline Conveyor Belt and Ice Ages

It has been the general consensus of geophysicist, since the Milankovitch theory was introduced, that extraterrestrial, astronomical forcing produces the recent ice age cycles since about 2.4 Ma BP. These are variations in the Solar irradiance from slight variations in the elliptical orbit (eccentricity), a variation in the Earths tilt to its ecliptic (obliquity or inclination) and a rotation of the elliptical orbit (precession). It is now known to good accuracy that their cycles are 100,000 years, 41-43,000 years and 23-26,000 years, respectively, but it is not understood how they interact and some astrophysicists believe there are other cyclic periods affecting our climate. These other cyclic periods are evidenced by minor climate variations in the cold glacial periods and the variation in the Interglacial climates as depicted in Figures 49 and 51 for glacial periods and significant variations in interglacial periods in Figure 47. We present, as Figure 46, an example of how they can interact with resonances and destructive interactions. We can see that Earth has gradually experienced cooling as can be seen in the Lisiecki and Raymo (2005) data back to 500 Ma BP in Figure 36, Panel A. The downward trend in global temperature appeared to have been spasmodic from 500 to about 45 Ma BP in panels A and B, when then some infrequent cyclic oscillations began to occur. The current major cooling trend seems to have begun about 18 Ma BP but from 18 Ma BP to about 8 Ma BP there were few oscillations. From 8 Ma BP to about 1.9 Ma BP in Panel C of Figure 36, we see that Earth experienced a cooling period but still intermittent oscillations. Between 2.9 and 1.9 Ma BP, as the cooling trend continued, these infrequent cycles began to become larger and consistent with amplitudes on the order of 5 °C producing the 43,000 year ice age cycles predicted by the Milankovitch theory. There was a temporal region between 1.9 and 1.1 Ma BP where the ice age cycles were somewhat chaotic. Beginning about 1.1 Ma BP, the dominant cycle began switching over to 100,000 years and has continued to the present.

We believe that the gradual cooling of the Earth's atmosphere can be explained by the movement of Earths continents, the evolution of the ocean circulation patterns and in particular the establishment of the North-South Thermohaline circulation system. Trying to go back further than about 100 Ma BP is difficult because the Earth's continents were floating around on Earth's surface with obvious effects on local land mass climates (see Scotese 2003), for example during the Carboniferous Era where coal was formed in West Virginia, Europe and the Urals which were together with tropical climates on the Equator and with swampy terrain. We believe the Earth began cooling when Pangea began breaking up about 500 Ma BP. Before with only one land mass near the Equator, there was minimum ocean circulation. By about 80 Ma BP, the Atlantic Ocean was opening up by the tectonic division of Europe and North America and between Africa and South America from the super Pangea continent. Antarctica was rifting from Australia and positioning itself at the South Pole and the Arctic Ocean was being closed in by Alaska, Canada, Greenland, Northern Europe and Russia.

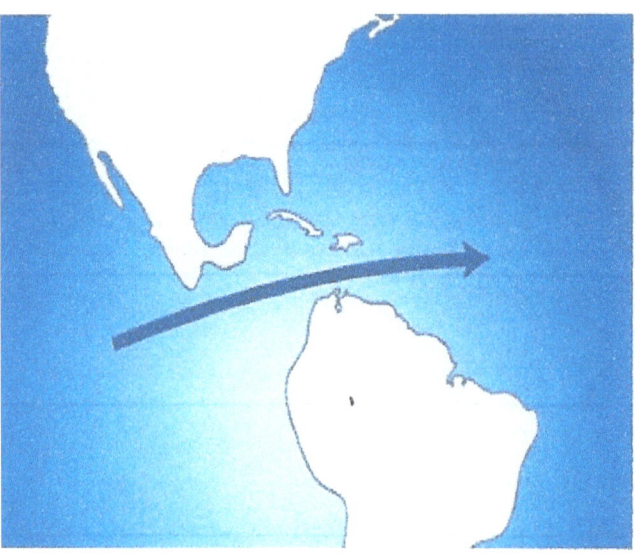

Figure 81, Ocean Passage between Pacific and Atlantic and Caribbean Sea Through Atrato Seaway in Panama.

  Even with North America drifted away from Europe and South America drifted away from Africa, there still existed an East-West ocean circulation between the Atlantic and Pacific through what is called the Atrato Seaway, shown in Figure 81. With the Atrato Seaway there was an exchange of tropical water masses between the Pacific Ocean and the tropical Caribbean Sea, Gulf of Mexico and Mid-Atlantic Ocean, the mixing provided a reduction in salinity and density. But the Seaway was gradually closing during the period from about 60 Ma BP by the collision of the Panamanian peninsula with the then separate South American continent. By 20 Ma BP the Seaway was said to be only several hundred kilometers wide (Kirby et al 2008), but still allowing the warm Caribbean waters to outlet into the Pacific Ocean. It is recently estimated that final closure of the seaway and termination of the East-West ocean circulation occurred about 4 Ma BP (Kirby et al 2008). The closure of the Panama Isthmus had a fundamental effect on Global ocean circulation, and led to profound changes in global climate by the Solar evaporation and the Caribbean waters high salinity and density, strengthening the Gulf Stream and activating the thermohaline downwelling (from the high salinity and density) in the North Atlantic around Greenland and Iceland. This offered warm climates for North America and Europe but also this facilitated the advent of the ice age cycles by disruption of this new thermohaline circulation by small extraterrestrial Milankovitch type Earth orbital oscillations. Figure 82 provides a sketch of the Caribbean region showing the closure of the Culebra Straits and just before closure of the Atrato Seaway 15 million years ago.

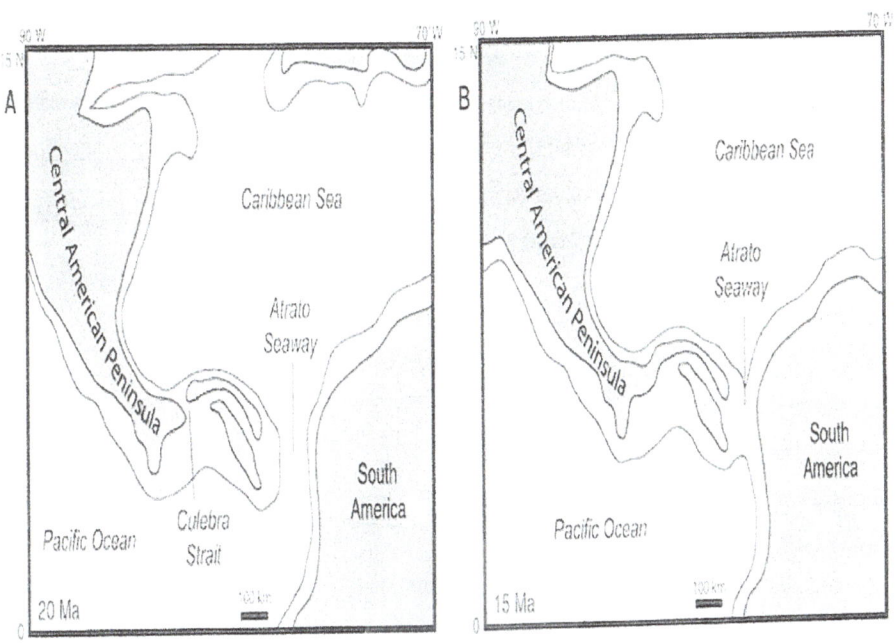

Figure 82, The Closure of the Atrato Seaway by Plate Tectonics

As the tectonic plate movement joined the peninsula with South America to form the Isthmus of Panama, equatorial ocean currents between the Atlantic and Pacific were cut off, forcing water northward into and strengthening the Gulf Stream. Figure 83 cites the effects of closure of the Atrato Seaway.

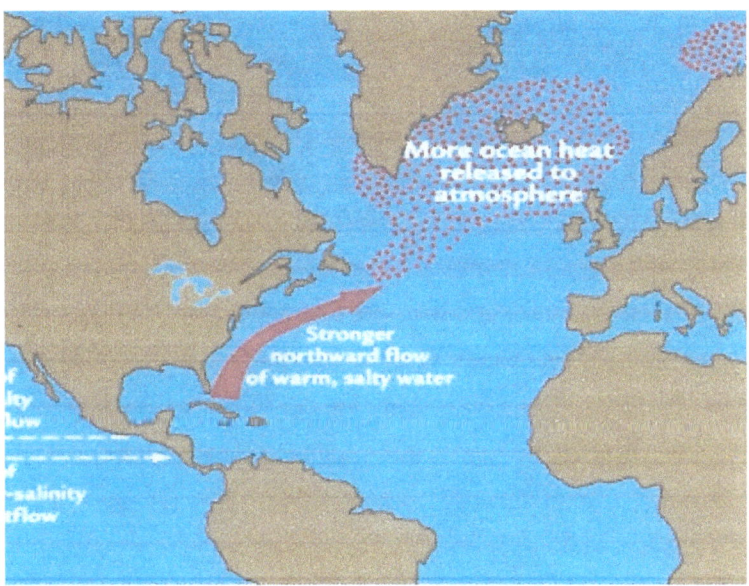

Figure 83, Pacific and North Atlantic effects from Closure of the Atrato Seaway

105

The North-South orientation and the westward movement of the Western Hemisphere continents is unique in Earths long 4.6 billion year history. The westward movement is as if the Earth is trying to balance itself, as one balances the tires on their automobile to prevent a wobble and accelerated uneven wear. The result is, not just an Atlantic Ocean circulation system but a global system of ocean currents called the Thermohaline Conveyor Belt. We show this World-wide circulation as Figure 84.

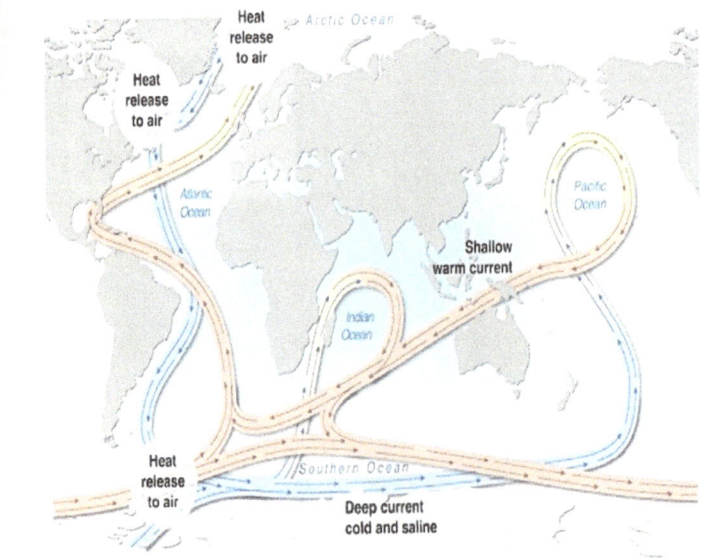

Figure 84, World-Wide Ocean Currents After Closure of Atrato Seaway

As salty water moves north in the Atlantic – carried by the Gulf Stream – it gets colder, as shown in Figure 85.

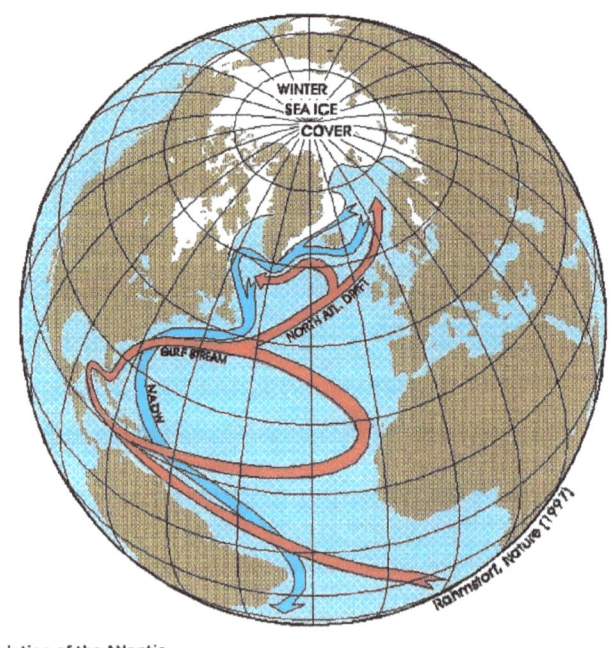

Figure 85, Atlantic Ocean Northern Currents

106

The combination of extra saltiness and cold temperatures makes the water more dense and especially proned to sinking. In the vicinity of Greenland and Iceland the salty, more dense water sinks to the ocean floor. Raymo et al (2004) have studied the Thernohaline effect on climate from sea floor cores in the North Atlantic finding that between Greenland and Europe the Gulf Stream waters are cooled and sinks producing an ocean floor counterflow Southward as shown in Figure 86.

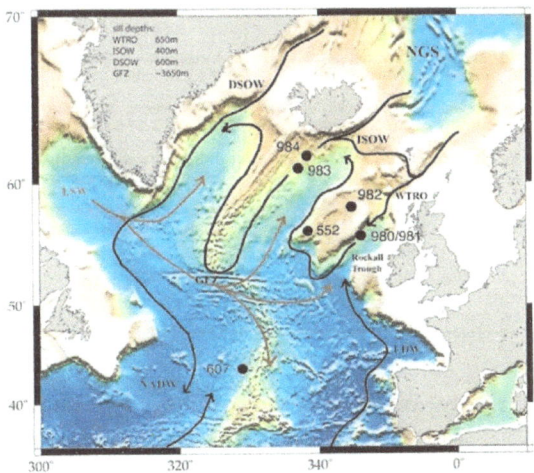

Locations of DSDP and ODP cores used in this study. Paths of major deep water flows are indicated by arrows.

Figure 86, Thermohaline Flow Patterns and Locations of Deep Ocean Drillings

From there it continues to spread out southward all the way to Antarctica, converges with another sinking current, and loops through the Indian Ocean and into the Pacific. There the water wells back up to the surface and slowly returns to the Atlantic, over top of the opposite current, around the tips of South America and Africa. The entire Conveyor Belt is driven by the sinking of the massive volume of salty and cooling water in the North Atlantic that was facilitated by the unique closure of the Atrato Seaway about 15 million years ago and strengthening of the Gulf Stream. Therefore the salinity difference between the Atlantic and Pacific caused by the Sahara Desert generated mid-Atlantic trade winds, that are thirsty for moisture, cause significant ocean evaporation, leaving behind a salty sea and dumping the fresh water as rain in the Pacific west of the Isthmus of Panama. With the enclosure of the Arctic Ocean, the Conveyor Belt no longer warmed the Arctic and sea ice formed the Arctic ice cap. Rather conclusive evidence that the Conveyor Belt became inoperative or at least much moderated during the past ice ages is a look at the geographical shape of the North Carolina and South Carolina Atlantic coast shore lines.

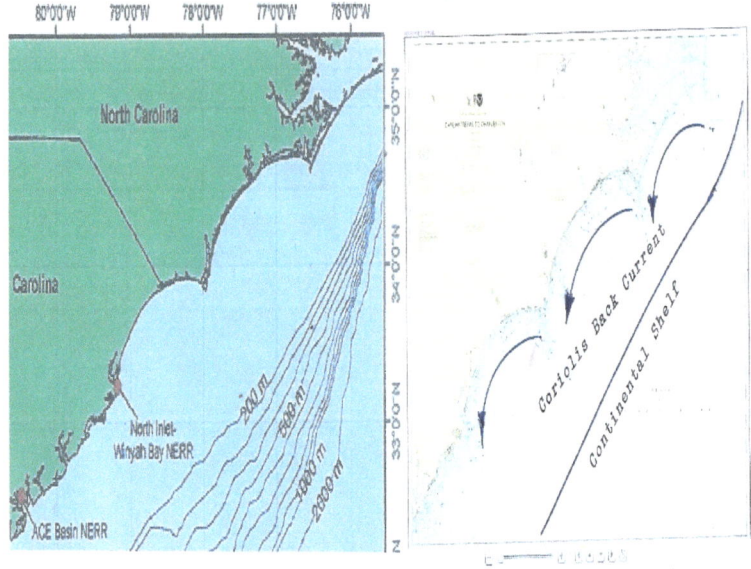

Figure 87, Left panel – East coast of US showing ocean floor terrain and the continental shelf ice age land mass. Right panel – NOAA chart showing ocean current eroded South and North Carolina con-cave shore line from Coriolis back current effect.

The Figure 87 provides, in the left panel, coast line and the ocean floor terrain of the region. The right panel provides the NOAA chart for this coastal region. Noted on the right panel NOAA chart is the edge of the continental shelf which is seen, in both the left and right panels, to be a uniform, regular shaped shelf continuing from Southwest to Northeast all the way past Virginia, Maryland and Delaware. The ocean floor between the present shore line of North and South Carolina and the shelf edge was dry land during most of the past ice ages. We next call the reader's attention to the present North Carolina and South Carolina shore lines. The three concave shaped coast lines are very conspicuous. Because the Gulf Stream, originating in the Caribbean Sea and Gulf of Mexico as shown in Figures 83 and 84, flows Northward past Florida as the Florida Current and Georgia, South Carolina and North Carolina as the Gulf Stream. In the middle of this Conveyor Belt, NOAA data reports a Northward ocean current of up to 5.5 knots (1 knot = 1 nautical mile per hour = 1.15 statute miles per hour) and a volume flow of up to 80 million cubic meters per second. On the surface of the Earth, as a result of the Earth's rotation, objects moving in the Northern Hemisphere experience a force to the left called the Coriolis force. This is why low pressure weather systems, hurricanes and tornadoes have a counter-clockwise rotation in the Northern hemisphere (and clockwise in the Southern hemisphere). A simple test is to fill your sink with water and watch the water disappear down the drain. It always rotates counter-clockwise – always here in the US and 39the Northern Hemisphere. Then the Gulf Stream Northward flow experiences some Coriolis influence and a counter-clockwise circular flow is generated to the left of the main flow, creating the Southward back current along the East coast as shown in the Figure 87. It is common knowledge that the U. S. Coast Guard's advice for low speed vessels cruising South is to either stay close in to shore, taking advantage of this reverse back current or go directly across perpendicular to the Gulf Stream and proceed South outside the Eastern edge. During the interglacial periods this back current has eroded the shore line and produced the concave coast line. This is true more so along the South Carolina and North Carolina coast because the Gulf Stream is closer to shore there since it is narrowed in flowing past the Bahamas. This erosion is interrupted by the presence of the Cape Fear and Savanah Rivers, their sediment causing three concave shore lines. Lynch-Stieglitz et al (1999) provide data showing that the Gulf Stream flow through the Florida Straits to be much weaker during the last ice age when the shore line was on the edge of the continental shelf. Thus, the shore line would not experience the back current erosion. The stream diverges Eastward past Virginia and further North, reflecting no Coriolis shore erosion.

We will show in later sections that the disruption of this Conveyor Belt has caused major climate cooling in Europe and North America but even observed in Antarctic ice cores. This most probably initiated the ice ages which began about that time, but without the Gulf Stream, North America and Europe would be permanently colder and perhaps no Interglacial warm periods. Closure of the Atrato Seaway is when significant Glacial-to Interglacial-back to-Glacial cycles began. But without the Conveyor Belt, the Northern Hemisphere would be much colder and perhaps a Snowball Earth in the future. The cyclic variation of the extraterrestrial forces caused the cyclic cooling and the sub session of the Conveyor Belt and onset of the ice ages.

Formation of Antarctica Ice Sheets

From global sea level data it is possible to trace the formation of large glaciation and ice caps. As noted Antarctica began separating from Australia about 100 Ma and by 90 Ma BP was in the extreme southern latitudes. With sea level analysis, Miller et al (2005) shows that about 90 MA BP glaciation began in Antarctica causing a 15 m global sea level fall and two Antarctic glacials between 2500 and 3000 m thick. By about 14 Ma BP, Antarctica had completely covered to its ocean edges with the present permanent ice cap (see their Figure 2) caused a sea level decline of about 63 m.

So the plate tectonics are responsible for 1.) the closure of the Atrato Seaway connecting the Atlantic and Pacific Oceans, 2.) the blocking East-West ocean currents, 3.) the creation of the enclosed Arctic Ocean to form the Arctic Ice Cap and 4.) creating a large South Pole land mass to accumulate glacial ice sheets, thus a partitioning of global temperatures between the two poles and the Equatorial tropics. It is like the old fashion ice boxes where the ice man would regularly deliver blocks of ice to your door and by putting the block at the bottom, the upper food storage region would be cooled by convection. Some ice boxes had ice at the top for better convection. Since about 20 Ma BP, with the closure of the Atrato Seaway, the enclosure of the Arctic Ocean and the positioning of Antarctica at the South Pole, the presence of these ice masses, with increased albedo, has reduced the net amount of Solar heat retained by Earth and we would expect lower global temperatures as Figure 49 shows. As we have noted, land masses more easily form ice sheets by net retention of snow in between seasons. The Northern Hemisphere has the majority of the Earths land mass and is therefore more susceptible to ice sheet formation. Earth has thus become a planet that is very fragile with respect to ice ages especially when changes in the Northern Hemisphere's Solar irradiance happens. The positive feedback of increased albedo, from increased ice cover exacerbates our planets fragile condition. From the closure of the Atrato Seaway by the Isthmus of Panama and the formation of the Thermohaline circulation, the Northern Hemisphere is warmed as long as this thermal Conveyor Belt is fully operative. We have seen on several recent occasions, the Younger Dryas and 8.2 ka Events, that just from a small amount of cold glacial water mixing can disrupt the Thermohaline circulation and cause significant cold spells.

Although triggered by extraterrestrial forces, the cause of the abruptness must lie in the internal dynamics of the ocean itself. Some have suggested that the ocean circulation in the Atlantic has two quasi-stable modes (Kobashi et al 2008, Rahmstorf 2002). During the glacial period, the ocean is in the cold stable mode with little or no Thermohaline circulation. During the Interglacial period, the ocean is in a quasi-stable warm mode with the Thermohaline Conveyor Belt in full flow. We use the prefix "quasi" since we have seen that cold water flooding can disrupt the Thermohaline circulation and temporarily "trigger" the cold mode. It seems though that during the Interglacial Plateau Region the extraterrestrial forcing has been strong enough to restore the Thermohaline circulation and the warm ocean mode a large portion by the warming of the waters in the Gulf of Mexico and the Northward flow of the Gulf Stream. We have shown with Figure 77, that similar disruption have occurred at the beginnings of the Interglacial Plateau Regions of 130 ka and 330 ka cycles, we suggest this to be apparently from cold fresh water flooding.

We are somewhat comforted in realizing that these events did not trigger a return to global glacial ice age conditions, however this appears to have happened for the 230 ka cycle where no definable Interglacial Plateau occurred (see Figure 63). General agreement exists that Earth will most certainly experience another ice age (a Las Vegas casino would give very high odds in favor), the basic question is when? We have noted that some investigators have premised that continued global warming could cause enough cold melt water production from Greenland and the Arctic ice cap to again disrupt the Thermohaline Circulation and pre-maturely trigger our next ice age within a few years' time (Barry 2005). Then there are others that premise that CO_2 global warming could delay this in-evitable ice age (Tyrrell 2007).

The Earths thermal status is indeed very fragile with the Arctic Ocean enclosed and Antarctica located at the South Pole. For millions and even billions of years in the past, Earth has enjoyed a semi-tropical climate in periods of relatively high airborne CO_2 concentrations, from Berner and Kothavala (2001) mean values ranging from present 200-360 ppmv to 1400 ppmv, and temperatures from 20 °C to 50 °C, without 23,000 year, 43,000 year or 100,000 year periodic ice ages millions of years ago [see Figure 76, Pagani et al (2005)]. The extraterrestrial, astronomical Earth orbit forces are weak but since Earths continent configuration has, and will be as it is at least for millions of years into the future until we drift to a better configuration, placing us in ice age jeopardy. It seems also that the three astronomical cycles may have entered into a harmonic resonance. We must keep in mind that the positive feedbacks, such as ocean CO_2 solubility decrease with increased temperature and decreased albedo with ice cap melting's, assists the weak astronomical forces. We premise this resonance because there were no consistent cycles until about 2.9 Ma BP, then a 23,000 year cycle and small amplitudes, the a 43,000 year ice age cycle with little larger amplitudes and now 100,000 year cycles with large amplitudes that seem to be getting larger. This is what would be expected for multiple cycles being out-of-phase and over time gradually coming into in-phase resonance. We have shown that the Glacial-to-Interglacial-back to-Glacial cycles seem to have a 200,000 years component which would be from multiples of the 43,000 year period. This phase resonance is suggested by Tzipermam et al (2006), who refer to it as non-linear phase locking for our present 100,000 year cycles. So can we expect the cycles to progress to an out-of-phase condition about 2.9 Ma into the present – if indeed we are at the maximum resonance in our present 100,000 year cycle reinforced with the 43,000 and 23,000 year cycles (and other cycles that we may not know about). Then can we expect a waning of the 100,000 year dominance and a resumption of the 41,000 year cycles at that time? Then without major cycles can we expect the cold trend of Figure 36 to continue to a colder and colder Earth? We can only speculate.

So what are the answers? Before about 15 Ma BP, the data of Lisiecki and Raymo (2005), Figure 36, shows only spasmodic occurrences of climate cycles. This was before the Atlantic Ocean had been formed and closed into a North-South ocean circulation configuration by the joining of the Isthmus of Panama with South American. From their data, we have averaged the temperatures over 200,000 year periods and provide this as Figure 36D. We see that the average Earth temperature has decreased from about 7 °C to about -5 °C. We have attributed this to the closure of the Atrato Seaway joining the Americas, the movement of Antarctica to the South Pole and the Quasi-enclosure of the Arctic Ocean creating an "ice-box" effect. What we may expect, with -5 °C continuous temperatures after the out-of-phase subsiding of the astronomical forcing resonance, is a cold Earth with semi-permanent North American Laurentide, Northern Europe and Asian ice sheets with-out any Interglacial warming.

Chapter 30 - A Mimic of Interglacial Warm Periods in the Past 500 ka BP

Relative to the three Malankovitch cycles shown in Figure 45, the last four cycles indeed reflect their strong influence. The 100 ka resonances are predicted with eccentricity providing additional influence each 400 ka periods. The EPICA data shows the 100 ka cycle influence on the past interglacial at 130 ka BP and the two interglacials prior to the 130 ka BP. We have performed a covariance test on two of these as shown in Figure 88. These are the MIS 5, 130 ka BP and the MIS 9, 320 ka BP interglacial periods.

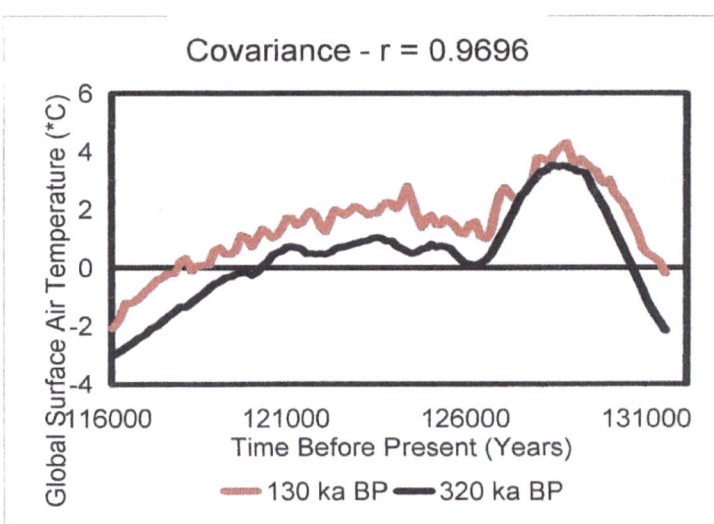

Figure 88, Graphics for Covariance and Correlation Coefficient Calculation Between the MIS 5, 130 ka BP and the MIS 9, 320 ka BP Interglacial Periods.

The covariance test yielded a value of 2.109 for the Covariance and the resulting Correlation Coefficient was a nearly perfect r = 0.9696, indicating a nearly identical astronomical forcing for these two interglacials spaced in time 200 ka apart. The range of times for ice core samples for the correlation coefficient calculation were 131,425 to 116,033 BP for MIS 5, 130 ka BP set and 336,425 to 321,033 BP for the MIS 9, 320 ka BP set for a total interglacial plateau of 15,392 years for both. The shapes of the two interglacials are nearly identical but the mean temperature of the 130 ka BP interglacial is about 1.05 °C higher. These Covariance results reveals how the astronomical forces can replicate themselves with many, many thousands of years in between.

The "Near-Time" Dominant Climate Factors – A Mimic of 430 ka BP Interglacial?

We next examine the two interglacial period that suggest a 400 ka astronomical resonance i.e. the MIS 11, 430 ka BP interglacial and our present Holocene interglacial. In particular, we will statistically examine how closely the two interglacials duplicate each other.

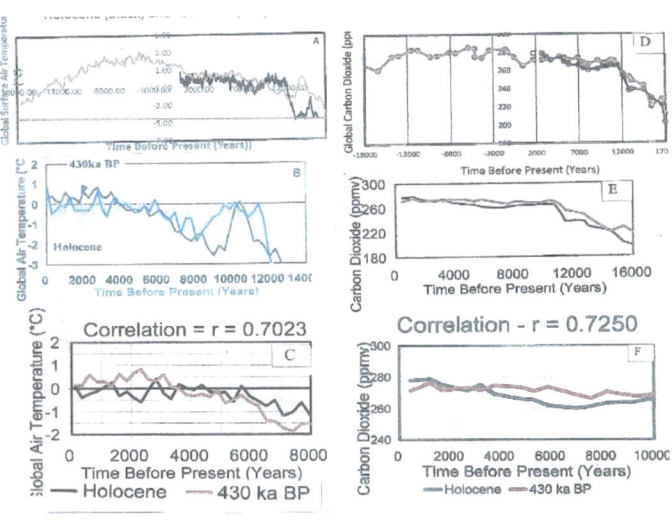

Figure 89, Correlation Between the MIS 11, 430 ka BP Interglacial and Our Present MIS 1, Holocene Interglacial for Both Global Surface Air Temperature and Atmospheric Carbon Dioxide.

Figure 89, Panels A and D show the two interglacials where the terminus regions are matched, being a smoother version of A. Panels B and E show the expanded Terminus and Interglacial regions and Panels C and F provide the Interglacial regions where the covariance comparison was calculated.

Panels B and E of Figure 89 provides the two interglacial temperature and carbon dioxide data covering only the regions to the present, the black curve being the Holocene EPICA data and the red curve the 430 ka BP interglacial EPICA data. It is seen that the beginning of the 430 ka BP plateau apparently experienced a cold event at very nearly the same location as the 8.2 ka Event in the Holocene period. We can speculate and suggest that this could be from a North American Lake Agassiz release event. Due to this cold event reflected on the EPICA data, for the matching of the two interglacials we began the covariance analysis for temperature at 8.2 ka BP where the climate was more stable for the two Interglacial periods as seen in the two sets of EPICA data. The EPICA team took fewer data points for the 430 ka BP plateau. In the plateau beginning at 8.2 ka BP there were only 31 EPICA data points. Further, the intervals were needed to be matched for the covariance protocol, so interpolation was performed on the Holocene data. The temperature averages for the covariance regions were -0.083 and 1.048 °C for the 430 ka BP and Holocene, respectively, just 1.131 °C difference for interglacial periods 400,000 years apart. Panels C and F of Figure 89 shows the two matched sets of data for covariance. The covariance computations yielded a covariance of 0.2298 and a value of Pearson's correlation coefficient, r, of 0.7250 for temperature. Guidelines for the use of the correlation coefficient provide that values from 0.7 to 1.00 indicate a strong positive linear relationship via a firm linear rule. The temperature correlation value of r = 0.7250 then further confirms the astronomical connection between the two interglacials. Then, as now, the Earth is in its near circular orbit. These results provides strong evidence that our Holocene interglacial period most likely will duplicate the 430 ka interglacial period as others have premised and as MIS 9 was duplicated by MIS 5 shown in Figure 88. We suggest that our Holocene Interglacial period is not 2 ka too long as premised by others and the Holocene terminus should not be synchronized with the

113

later portion of the 430 ka BP interglacial as suggested by Rohling et al (2010), but with the actual initial terminus of MIS 11.

We can examine the similarity between the Holocene and 430 ka BP interglacial plateaus in more detail. In Figure 90, we show the EPICA data again but here we have added the NASA Goddard Institute for Space Studies (GISS) data with the base period as 1951 to 1980 AD shown in Figure 91. The 5 year mean values were used.

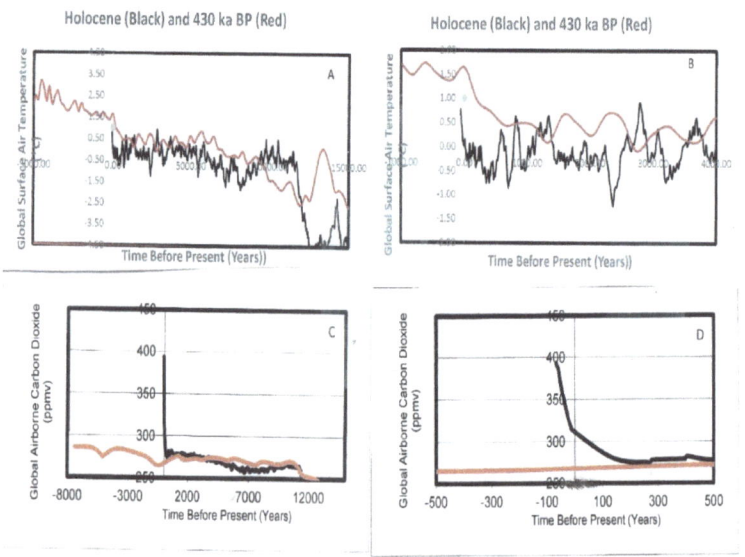

Figure 90, Excel graphs of EPICA temperature and Carbon Dioxide data with 1950 AD as Zero Time Before Present. The most recent NASA GISS for Temperature for 2013 is included. The red and black data are the 430 ka BP and Holocene EPICA data, respectively.

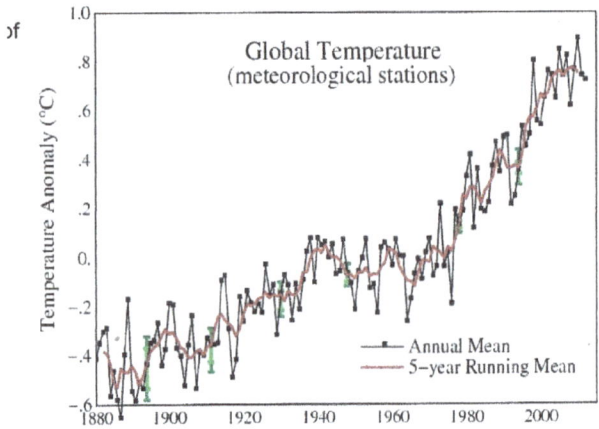

Figure 91, NASA GISS Global Temperature

What these graphs show in Panel A and B is that the MIS 11 430 ka BP experienced a sharp rise in Surface Air Temperature of about 1.3 °C about 15,000 years after the Terminus was completed. In our present Holocene period we are experiencing a similar temperature rise, seen in Figure 91 and in panels A and B of Figure 90. IPCC is attributing all of our current Holocene period temperature rise to man-inflicted carbon dioxide shown in Panels C and D. IPCC further attributes minimal Global Warming to Solar Irradiance forcing i.e. in FAR 2013, 0.05 versus 1.65 W/m² for carbon

114

dioxide. ICPP (2013) provides their FAR 2013 Radiative Forcing breakdown. But observing the abrupt increase, at about the same interglacial plateau time for the 430 ka BP period and because our Holocene so closely mimics this interglacial to within a few percent, it is not appropriate to credit the Figures 90 and 91 temperature rise solely to Global Warming from human CO_2 production. Similar abrupt rises in temperature have been experienced in the recent past both in the last glacial period from D-O and H Events and in the present Holocene period. Table 1 summarizes those in the Holocene period and past glaciation.

TABLE 1
ABRUPT TEMPERATURE CHANGES - Time to EPICA 1950 AD

| | Holocene Period | | Past Glacial | | 430 ka BP | |
	Time BP	Temperature °C	Time BP	Temperature °C	Time BP	Temperature °C
End	-60	0.77	27333	-6.91	411402	1.64
Begin	73	-0.42	27523	-10.51	412046	0.43
Difference	133	1.19	190	3.6	644	1.21
End	580	0.34	28419	-7.84	412277	0.49
Begin	701	-86	28658	-9.46	412739	0.08
Difference	121	1.2	239	1.62	462	0.41
End	812	0.62	31618	-10.36	412977	0.66
Begin	961	-0.49	32012	-6.9	413432	0.23
Difference	149	1.11	394	3.46	455	0.43
End	2236	0.04	32012	-6.9	413902	0.57
Begin	2366	-1.23	32761	-9.89	414139	-0.08
Difference	130	1.27	749	2.99	237	0.65
End	3074	0.37	37431	-8.14		
Begin	3207	-0.74	37819	-4.66		
Difference	133	1.11	388	3.48		

Table 4, A listing of occasions in the present Holocene period, the past glacial period and the 430 ka BP interglacial where relatively abrupt temperature changes occurred.

For example in the first listing in the Holocene column is the present rise in temperature shown in Figure 91 beginning in 1877 (73 years before 1950 AD) and ending with the EPICA data for 2010 AD (60 years after 1950 AD), a total time period of 133 years for completion of the change. The Holocene data are taken from the EPICA data shown in Panel B of Figure 90 and showing relatively abrupt temperature rises comparable to our present Figure 91 rise. The past glacial data show much larger abrupt changes due to the unexplained D-O and H Events but believed by some to be of astronomical origin. In the 430 ka BP interglacial the abrupt rise beginning at 412 ka BP coincides with our present Holocene abrupt rise but taking longer to complete. We have the Pagani et al. (2005) data to estimate how much of the temperature rise in Figure 91 is from human generated carbon dioxide and how much is astronomical forcing? During the period from 1880 to 2013 AD airborne CO_2 has increased from 280 to 400 ppmv. We show in Figures 76, Panel A and 76, Panel B the minimal increase in Surface Air Temperature for CO_2 increase from about 280 ppmv. Using the best fit to the Bode type feedback Equation (13), the increase in temperature from an increase on CO_2 from 280 to 400 ppmv was 0.636 °C. Then in Figure 90, our best estimate is that of the Global Temperature rise from about -0.5 to 0.8°C (change of 1.3 °C) only about half, 0.636 °C, should be attributed to CO_2. We conclude that carbon dioxide and other GHGs have not provided any significant influence in Global Climate in the past interglacials where human influence was absent. With Figures 76, 89 and 90, we further conclude that astronomical forces from the Milankovitch cycles in Figure 76 have dominated the past four interglacials and have totally dominated our pre-industrial part of our present Holocene interglacial periods.

115

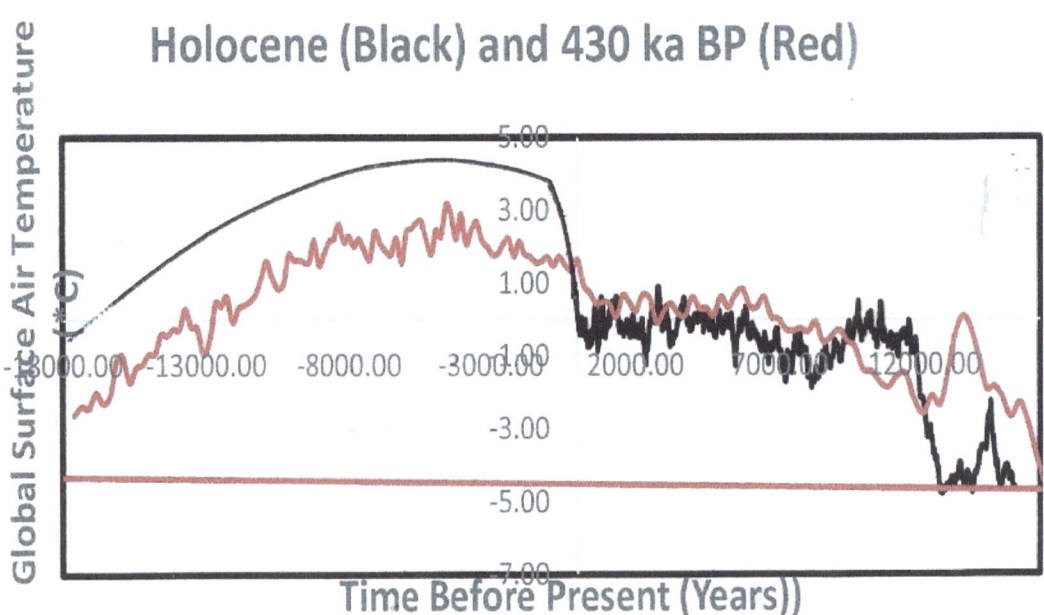

*Figure 92, Estimated Global Surface Air Temperature Through the Remainder of Our Present
Holocene Interglacial Period*

IPCC (2013), in their most recent fifth assessment report (AR5), have predicted future
Global CO_2 increases and corresponding radiative forcing values of 2.8, 4.5, 6.0 and 8.5 Watts per
meter^2. They premise that Global Atmospheric Carbon Dioxide levels can reach as high as 950
ppmv by the year 2100 AD if their most liberal Representative Concentration Pathway with radiative
forcing of +8.5 Watts per meter^2 is achieved. This is if the present human CO_2 "pollution" rate is
continued. Currently the radiative forcing is 2.87 (2012 AD value) compared to 1.712 in 1979 AD.
Again the best information available to access this prediction is the Pagani et al. data where CO_2
concentrations to 2195 ppmv were observed with the corresponding surface air temperatures also
determined. Again we refer to Figure 76, Panel A and Equation (4). We expect that the Global
Temperature will continue to mimic the 430 ka BP Interglacial climate forcings from the
astronomical influences but also expect some temperature influence from the human generated CO_2.
The average temperature in our Holocene period is about 1.1 °C lower than that for the 430 ka BP
interglacial for the same time periods, as seen in Figures 92 and 93. In Figure 92, we chose to
estimate that CO_2 levels will reach 1000 ppmv before environmental efforts abate Global Warming
trends. This we acknowledge in Figure 62 with a maximum temperature for the remainder of our
Holocene period. Based on the Pagani et al. data and these considerations, we show Global Surface
Air Temperatures through the remainder of the Holocene period to be about 2 °C above the 430 ka
BP interglacial temperatures. This will still mean that our Holocene interglacial maximum
temperatures will be below the four other high interglacial temperatures in Figure 62. This is a rise
from about 1.1 °C below the 430 ka BP level. As discussed above, this we must attribute, from Table
1, about 1.2 °C to astronomical forcing and, from the Pagani et al. Figure 76 and Equation (4), about
1.8 °C to human generated CO_2. This then suggests a maximum of about 1.9 °C above the 430 ka

116

BP levels for the second half of our interglacial.

For a considerable period of time, it is expected that the Global Surface Air Temperature will remain at least 3.5 °C above EPICA mean temperature for the past 1,000 years. As shown in Figure 62, last interglacial MIS 5 -130 ka BP exceeded and MIS 9 – 320 ka BP equaled this temperature.

Two Considerations Relative to Accuracy of This Study

What we provide here in this text is only what we consider to be our best estimate of our future Global Climate. There are two factors that need to be examined relative to the accuracy of the work reported here. One is the potential difference in the age of the trapped CO_2 and the age of the ice itself. Gas is trapped in the polar ice at depths of 50 – 120 meters and is therefore younger than the ice in which it is trapped. This has complicated the findings of Indermuhle et al. (2000), Hansen et al. (2007), Soon (2007) and Stott et al. (2007) indicating a lag in changing CO2 levels behind increasing global temperature change thus supporting the contention that Global Climate is dominated by temperature (thus astronomical) forcing. There has been considerable uncertainty as to the magnitude of this gas age to ice age difference, Δage, in evaluating ice core data for the temporal scale for past CO_2 concentrations relative to air temperature. Bender et al. (2006) estimate the uncertainty to be ± 1,000 years. Loulergue et al. suggests that the phase relationship between CO_2 and temperature phases is overestimated by Bender and others. To examine this with respect to our CO_2 and Global Temperature correlations between the MIS 11 (430 ka BP) and our interglacial, we have recalculated the correlations with a phase shift of 700 years and found minimal difference in the linear best fits. We show the Figure 93 for the 570 ka BP interglacial with the EPICA data in phase and with the blue line best linear fit with a 700 year phase shift. Two best linear fits are – In Phase T = 5.312 + 0.02748 x CO2 and 700 year out of phase – T = 3.660 - 0.02077 x CO2.

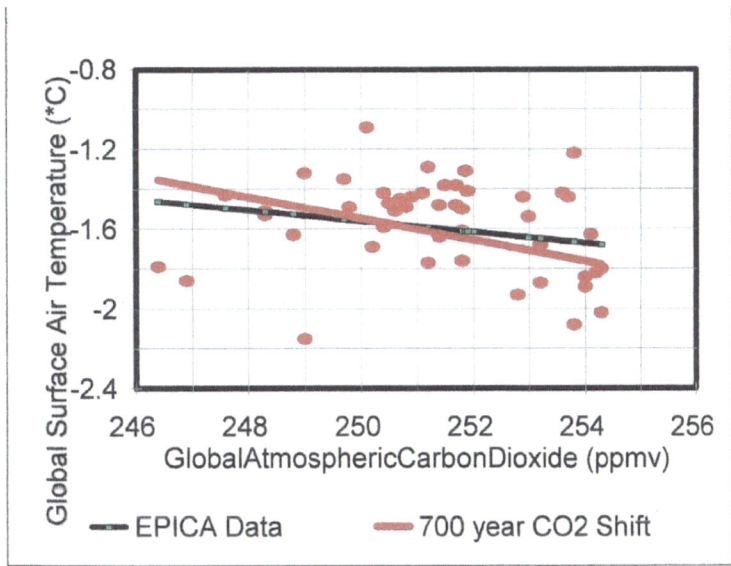

Figure 93, Linear Best Fit of 570 ka BP Interglacial Period to in Phase EPICA Dome C Data and Fit With 700 Year Out of Phase CO2

117

Figure 93 shows the in phase EPICA data best linear fit and the best linear fit with a 700 year shift for younger CO_2 in the ice cores. We conclude that the gas age – ice age difference does not affect our conclusion that CO_2 has not in the past and does not now dominate our Global Climate.

A second consideration is a question of the validity of applying the data of Pagani et al. (2005) to estimate the future temperature increase from the expected man generated increase in Global Atmospheric Carbon Dioxide Concentrations. The Earth's continent configuration at that time was similar to the present configuration (see Scotese 2003) with the Atlantic Meridional Overturning Circulation (AMOC) in operation. Without this ocean flow, the Pagani et al. temperatures could be an underestimate by a few degrees. However, Lynch-Stieglitz et al. (1999) found that during the last ice age the Gulf Stream only slowed by one third.

We in our Holocene Period are approaching a fully melted Greenland and Arctic ice cap. We should expect a similar Global Surface Air Temperature rise as happened 400 ka ago. We consider the Solar sunspot cycle and other short-term cyclic effects to be minor compared to the affects cited above although, with the fragile condition of Earth, we may experience "Little Ice Ages". Based on past Interglacials, without greenhouse effects, we expect all of the Greenland ice sheet may melt and the Arctic ice cap to seasonally disappear. Above we have acknowledged the recent report in Nature that the Northern portion of the Greenland ice sheet remained during the last Interglacial Period 120,000 years BP (Dahl-Jensen 2013), but the 430 ka BP, which we appear to be mimicking, was much longer in duration. Prior to their results it was thought that the Antarctic Ice Sheet may be little affected and perhaps even grow in volume from increased atmospheric moisture due to a warmer climate.

In this study we have examined the early evolution of the carbon dioxide and oxygen atmospheric components. We examine if all the oil deposits (230 Gt) on Earth were consumed and the resulting CO_2 remained in the atmosphere. This would add to the 720 Gt now in the atmosphere. The CO_2 content would then be about 470 ppmv, resulting in a global temperature increase of about 1.1 °C based on the data of Pagani et al (2005). This does not include future increases from cement manufacture and other human and animal contributions. Coal burning electric power plants are the largest human source of atmospheric CO_2. There is concern that we are reaching a "tipping point" where Earth will permanently have a run-away Greenhouse effect (Gore 2005, 2006, Hansen 2006). There are others that fear that the anthropic production of excess CO_2 above what Earth experienced in past interglacial periods may trigger a premature ice age.

DISCUSSION

The Feedback Effect of Atmospheric Carbon Dioxide with Respect to Global Surface Air Temperature

We have shown that Airborne Carbon Dioxide is not the driving force for the large temperature excursions during the past Glacial-to-Interglacial-back to-Glacial cycles (see Figures 68 through 74), but it is to the converse i.e. temperature warming the oceans and land and driving the CO_2 excursions in Figure 64. It is obvious that CO_2, methane and nitric oxide all create a greenhouse heating effect, as we pointed out with our neighbor Venus. We have estimated the magnitude with simple attenuation theory and have used the Ma BP data of Pagani et al to obtain an Earth derived attenuation constant, given by $T = To \, Exp(-\mu C)$, of $\mu = 9 \times 10^{-5}$ per CO_2 ppmv which is in good agreement with our value of 17.0×10^{-5} as shown in Figure 76.

The estimated relation between temperature and greenhouse CO_2 adopted by International Panel on Climate Change (IPCC) is $\Delta T = 6.8 \, Ln \, (C/Co)$ in the CO_2 range to 1000 ppmv. In Figure 76, Panel E we show the IPCC estimate, disagreeing considerably with our two estimates. As noted above, we estimate the CO_2 doubling temperature increase to be between 0.345 and 0.50 °C from CO_2. Above we have predicted a temperature increase of about 2.5 °C, not from GHG but from extraterrestrial forcing during the rest of the Holocene Period. Then perhaps a net of 3.0 °C including GHG as estimated in Figure 89.

Some Near Catastrophes for Our Atmosphere

Our goal of this text was to provide a concise documentation of the past history of our earth extending all the way back to the very, very beginning, the Big Bang. From this we have presented the present geo-physical status of earth and what the future might bring. To do this we have had to extensively address the present and potentially future global climate crisis.

Early in Earth's atmospheric history, the atmosphere was predominately carbon dioxide as was Venus and Mars. Essentially all of the carbon listed in Falkowski et al's Table 2 was in our atmosphere and but continuously being "scrubbed out" by precipitation. With the presence of liquid water, Berner and Kothavala (2001) have shown that the airborne CO_2 was removed from the atmosphere by precipitation, reacted with primarily Ca and Mg to form stable carbonates constituting a significant part of Earth's crust. The land (and ocean) deposited carbonate stable solids have then done two significant things, removed CO_2 from the atmosphere and also covered minerals that would react with the oxygen, giving O_2 a chance to accumulate in the atmosphere.

When flora life forms evolved, consuming CO_2 in the photosynthesis process while producing O_2, atmospheric CO_2 was further reduced. These two CO_2 removal processes are still very important in the present quasi-equilibrium status of our atmosphere. The oxygen produced in photosynthesis was first largely consumed by reaction with land surface exposed minerals such as iron and sulfur. About 95% of all Earth oxygen is estimated to be bound in these minerals. If more minerals were present in the Earths exposed crust, perhaps there would be no atmospheric O_2 today. Thus, Earth has avoided a major atmospheric disaster that could have occurred. Secondly, if there were no CO_2 producing fauna on Earth perhaps there would be much less atmospheric CO_2. The present CO_2, even with anthropogenic activities is at an extremely low level compared to the most prevalent level of over 4000 ppmv during the Ma BP years when life thrived, fossil fuels were being formed and animals, such as dinosaurs, heavily populated the Earth. We have pointed out that plant-life have had to re-configure their photosynthesis to the C4 pathway because of the drastic reduction in CO_2 concentrations to the present level. Without CO_2 producing fauna and with the continued locking-up of CO_2 into carbonates, flora life could have slowly vanished as CO_2 disappeared by carbonate weathering and its own CO_2 consumption. So Earth's atmosphere has then avoided two potential catastrophes. It is estimated that humans could survive with CO_2 levels as high as 1 % without respiration problems (10,000 ppmv).

Another potential catastrophe for Earth that has evidently occurred at least once in the past, before humans arrived, is the "Snowball" effect (Kirschvink and Kopp 2005), which is proposed to have occurred about 2.3 to 2.2 Ga BP during the Makganyene period. A strong positive feedback would be that as more and more of the Earth became covered with ice, its reflective nature meant Earth's albedo increased greatly cooling the Earth even more. It is reasoned that Earth had only one way to escape its Snowball condition, this from volcanic activity. With Earth completely frozen, there would be no rainfall to wash out the atmospheric CO_2 such that it would gradually accumulate in the atmosphere increasing the GHG component till finally Earth warmed enough to melt the Earth out of its Snowball state.

We are of course hypothesizing, but we offer one more catastrophe. If the 65 Ma BP asteroid had not struck Earth, as mankind evolved it would have had to co-exist with carnivorous Dinosaurs which would be a severe impediment and we humans may not have reached our present Earth domination status.

Consequences of a Next Ice Age

We have learned a great deal about this beautiful Earth that we live on. In examining what are the most sever and most probable events that could happen to our paradise, there are several that are considered possible. One of course is a thermo-nuclear war. Russia and the United States still have thousands of thermo-nuclear (hydrogen bomb) weapons, even enough to obliterate civilization as we know it, but not destroy the human race. With the drive by Islamic nations to obtain nuclear weapons (India and Pakistan have them) it is most probable that a few low yield bombs will be used in local conflicts perhaps by terrorists groups. Based on the Hiroshima and Nagasaki experience, hundreds of thousands may be killed, but world organizations should be expected to certainly curtail any chance of nuclear weapons becoming a common instrument in wars. Another global threat is the probability of a giant asteroid striking Earth, as they have many times in the past, including the devastating one about 60 Ma BP. Nearly every year one passes close enough to pass between the Earth and our Moon. The last significant one to impact was about 60 million years ago when dinosaurs were obliterated. An international group is on asteroid watch and the US is studying ways to divert such an asteroid having Earth as its destination (one being use of nuclear devices to deflect the asteroid).

Perhaps the most certain catastrophe in Earths future is our next ice age which could be 1,000, 5,000, or even 20,000 years into the future, based on Figure 89. We have no control over this. We can assume that ice sheets will be again formed in North America, Europe and Russia to thousands of meters thick. Figure 94 shows the ice sheets estimated from the last ice age.

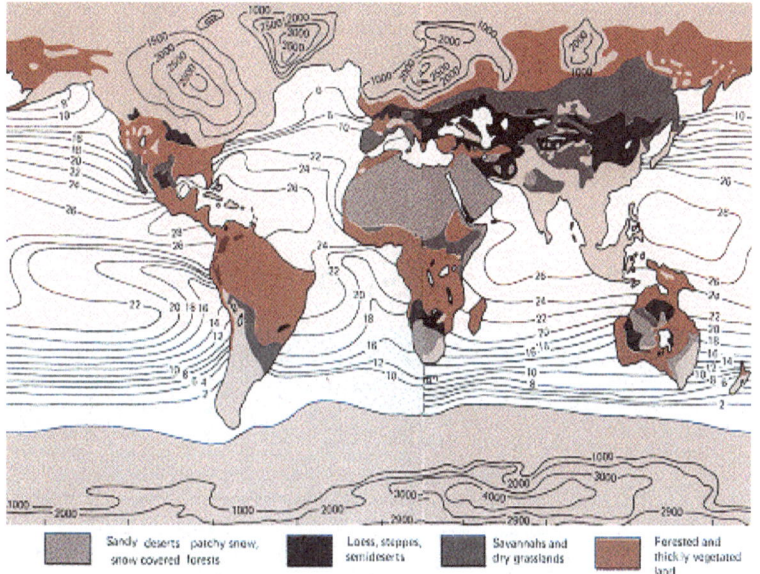

Figure 94, Ice Age Ice Caps Showing their Thickness in Meters

We see that North America had by far the largest ice sheet. However Europe now has the largest population density. We estimate that about 35 million persons will be displaced in Canada and another 50 million in New York, Pennsylvania, Michigan, Minnesota, Wisconsin and Alaska. In Europe, 60 million in the United Kingdom, 20 million in Norway, Sweden and Finland, 20 million in France and Germany and 20 million in Russia. This totals 205 million dislocated persons from areas where the ice caps will be hundreds to thousands of meters thick and not live-able. It can be safely assumed that there will be plenty of time for humans to do this i.e. hundreds of years for this relocation to take place.

The major economic expense after the relocations will be the massive increase in energy demand, which we think may quadruple the present energy consumption needs from the present 150 petawatt-hours (1.5×10^{17} watthours). The Ice Age will certainly disrupt and reduce the energy production methods that we have now, so it is apparent that there will be great hardships while enduring tens of thousands of years of ice age after its onset. Perhaps fusion energy will by then be harnessed (a gallon of water has enough deuterium if fully burned by fusion to produce the equivalent of 300 gallons of gasoline).

Correlation Between Global Temperature and Airborne Carbon Dioxide in all Ice Age Stages

By extensive analysis of the Antarctic EPICA site ice core data for both Global Temperature and Global Airborne Carbon Dioxide we have shown that CO_2 levels lag the temperature temporal variations during the Glacial-to-Interglacial-back to-Glacial cycles for the past 800,000 years before present. This is true for all phases of the cycles i.e. the Terminus Rise Region where the climate is rapidly coming out of the ice age, the Interglacial Plateau Region where the CO_2 continues to gradually increase and the Interglacial Fade Region where the temperature begins to return to the next ice age the CO_2 remains in the quasi-steady-state Interglacial Plateau phase for a longer period, as can be seen in Figures 63 and 64. This confirms then that temperature change is the driving force for the CO_2 global changes, apparently by warming the ocean and land mass temperatures and increasing CO_2 aspiration (CO_2 solubility in water deceases with temperature increase) to the atmosphere during the Terminus Rise Region. Similarly, as temperature decreases, the solubility increases as the air temperature cools the oceans and land. The Sea Surface Temperature is expected to lag the increased air temperature, and hence carbon dioxide expulsion from the oceans should lag temperature. Both are found to be true in the past.

We have followed the evolution of Earth from the Big Bang to our present state. We have undertaken an extensive analysis of the Earth's time behavior including the development of our present atmosphere and our Global Climate, examining all the parameters believed to be crucial to climate behavior. After reaching that point in the evolution of Earth, it seemed necessary to deal with Earth's current major problem, the obvious Global Warming that appears unprecedented since the advent of modern man. Following the logic in the other aspects of Earth's evolution, relative to Global Warming, we turn to the past to asset our present and future. From this study of the past, we make the follows conclusions:

- Although there has been 13.77 billion years since the Big Bang produced the initial protons, the universe is still primarily hydrogen atoms, all which were created at that moment.

- We have traced our Solar system and found it to be in the Local Interstellar Cloud of the Local Spur of the Orion Arm in our galaxy. This is on the outer edge of our galaxy about 8.33 kpc (27,000 light years) from our black hole galaxy center.

- The Moon most likely was created by collision of Earth with a Giant Impact Object that was accreted, we estimate in near-Earth orbit, within about 30 M km of our orbit with the Sun. Here we estimate that it took about 18 million years, from mutual gravitational attraction, for the collision to occur. It most likely was a near head-on collision (since their relative orbital velocities were very small) with a fast-spinning Earth based on Gott (2011) and Cuk and Stewart (2012), with the Moon being primarily Earth composition and Earth isotopic signature.

- Even during our present Holocene intergalactic warm period, the past Interglacial Periods minimal correlation between CO_2 and Temperature indicate that at least 90% of our climate is dominated by extraterrestrial forces. Airborne CO_2 concentrations have little temperature (and thus climate) effect.

- The closure of the Atrato Seaway in Panama, the positioning of Antarctica at the South Pole and the enclosure of the Arctic Ocean by Alaska, Canada, Greenland, Northern Europe and Russia has created a North Atlantic Thermohaline flow pattern, warming North America and Europe, but a unique condition in the history of Earth, where Earth has become very fragile to small changes in Solar energy transfer to its surface.

- The large variation in Global Surface Air Temperature during the glacial-to Interglacial-back to-glacial cycles is driven by extraterrestrial forces, which could be astronomical forces related to variations in the celestial mechanical properties of it rotation around the Sun such as eccentricity, obliquity and elliptical precession (Milankovitch cycles).

- Major climate fluctuations, far exceeding other Earth-bound perturbations from other sources, has occurred from the sudden release of cold water from Northern Hemisphere Glacial Lakes during Ice Sheet melting for Interglacial cycles with high Global Temperature rise as have been observed during our present Interglacial period and the MIS 5, 130 ka and MIS 9, 330 ka Interglacial cycles. This cold water flooding is believed to interrupt the Thermohaline North Atlantic Ocean circulation, reducing the transport of thermal energy to the Northern latitudes and causing abrupt cold spells such as the Younger Dryas and 8.2 ka Events and long duration Ice Ages. Again showing fragility of Earth's climate.

- In Figure 87 examining the Eastern US Atlantic Ocean present shore line and the continental shelf shore line that existed during the ice ages, it is conclusive that the Thermohaline Conveyor Belt was inoperative or drastically slowed during the past ice ages (thus inducing the ice ages). This is because Coriolis force erosion seems to have only occurred during interglacial periods from the Gulf Stream Northward currents on the present shore lines.

- We show evidence that Younger Dryas type glacial cold water flooding occurred at the beginning of the Interglacial Plateau Regions of the Interglacial cycles MIS 5, 130 ka and MIS 9, 330 ka. This is premised due to the high Interglacial Plateau Period temperatures shown in Figures 63 and 64 and the shape of their Interglacial Plateau's in Figure 64.

- Earth's variation in carbon dioxide concentration appears to vary independent of Global Surface Air Temperature during quasi-equilibrium periods such as during the Interglacial Plateaus (see Figure 73).

- During the Glacial-to-Interglacial-back to-Glacial cycles, the large increases in CO_2 levels during the Terminus Rise Region are driven by the increasing Global Temperature primarily from rise in Sea Surface and Earth Soil Temperatures and the resulting increase in CO_2 aspiration to the atmosphere. A lag time on the order of 1000 years is reported to occur from the observed lag times in Sea Surface Temperature and Earth Soil warming. The main source of the CO_2 is the large reservoir in the ocean (shown in Table 2), since CO_2 solubility decreases with increasing temperature.

- During the Interglacial Plateau Region of the cycle, the Global CO_2 concentration continues to increase due to this homeostatic lag effect of Sea Surface Temperature warming. This can be seen in Figures 63 and 64.

- There are very large variations in the slope and duration of the Interglacial Plateau Region of the temperature and CO_2 cycles. For the temperature, which is believed to be directly driven by the extraterrestrial forcing mechanisms by Solar Irradiance variation, this must mean that there are harmonic resonances and dampening's with the astronomical forcing cycles and one single forcing mechanisms is not always dominant, otherwise the cycle Plateaus duration and slopes would be more consistent if it were.

- In comparing the Plateaus of past cycles in Figure 65, the 430,000 year BP cycle provides significant similarity to Earths present interglacial cycle Terminus Rise and Interglacial Plateau Regions without considering any astronomical forcing theories. This has been observed and studied by others (Rohling et al 2010, Raymo and Mitrovica 2012, McManus et al 2002). No other past cycle offers any reasonable resemblance. Using Covariance analysis methods, we show in Figure 89 that the 430 ka BP interglacial and our present Holocene period are strongly correlated. From this analogy, we estimate that the current cycles Interglacial Plateau may last at least another 7,500 years. Based on prior, most recent Interglacial cycles, we should expect a 2.5-3.5 °C increase over 1950-1980 temperatures, the melting of most of the Greenland ice sheet and the Arctic Ocean ice cap seasonally and a global sea level rise of about 10 to 20 meters. This is irrespective of any CO_2 greenhouse effects, which are expected to be relatively small during this extended interglacial Plateau period.

- Based on modeling of CO_2 effects on Global Temperature and the data of Pagani et al (2005) of Earth CO_2 concentrations and Earth temperature over a CO_2 range from 360 to 1800 ppmv, we estimate a Global Warming coefficient of 0.09°C rise per CO_2 100 ppmv increase and a temperature increase rate of 0.342 °C per doubling of present CO_2 concentration.

- Some concern is expressed that fully melted Greenland and West Antarctic Ice Sheets may induce a premature ice age ("hosing"). Recent data show that they were most likely melted during the 430 ka BP Interglacial Period (Raymo and Mitrovica 2012) and should not induce an ice age in our present Interglacial.

APPENDICES

Parallax and Doppler (Red Shift) Methods in Cosmology Parallax Method for Planet, Stellar and Galactic Distance Determination

Parallax observations are the observation of the apparent change in direction of an object caused by a change in the observers position that provides a new line of sight. The term Parallax is derived from the fact that the lines of observation are very nearly parallel if the distances are much greater than the distance between the two observation positions.

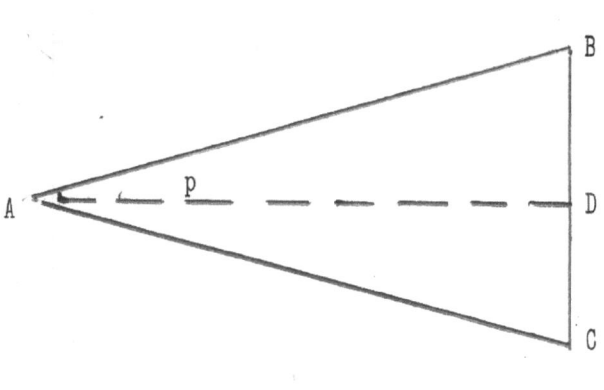

Figure A1 – Parallax of an Object, A, at a Great Distance Compared to the Distance Between Two Observations, Point A and B.

In Figure A1, we show the object under observation, A, and two observation points, B and C, a known distance apart. Point D is the midpoint between B and C. By measuring the difference in the angles of observation, the apparent shift (angular), p, is determined. By simple trigonometry for right triangles, the tangent of p equals the ratio of the opposite side (labeled BD) to the length of the adjacent side (labeled AD). For small angles (which this is for cosmological objects at great distances), the MacLaurin series expansion of the tangent, i.e. Tan (p) $= p + p^3/3 + 2p^5/15 \ldots$, gives the desired accuracy. For p less than $10° = 0.175$ radians, Tan (p) = p within ± 1 %.

Then

$$\text{Tan (p)} = (BD)/(AD) \approx p \hspace{4cm} A1$$

and solving for AD, we have

$$\text{Distance to object} = AD \approx BD / p. \hspace{3cm} A2$$

The observation points in cosmology could be the opposite sides of the Earth for Solar system measurements or the six months opposite positions of the Earth in its orbit around the Sun for Stellar or galaxy measurements. For the former, the base line, BD, would be about 6,378 km and for the later about 149.6 M km.

1. Einstein's Doppler or Redshift Effect For Cosmological Velocities and Distance

Einstein used the variation of the frequency (and thus wave length) of sound waves from a train passing by an observer to demonstrate the Doppler effect as it would also apply to cosmological light rays.

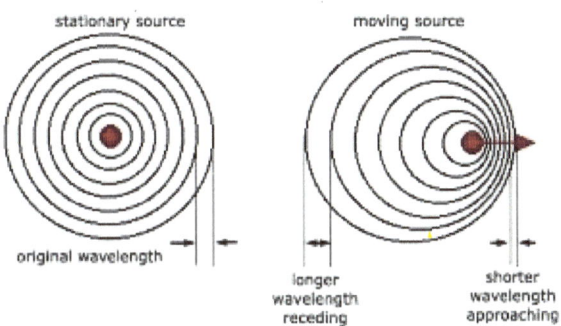

Figure A2 – The Lengthening of the Wave Length of Light Rays From Stars and Galaxies Receding From the Earth's Stationary Frame of Reference

For cosmological light rays, a Redshift occurs when a star or galaxy as a group of stars moves away from observers on Earth as a result of the Big Bang. A Blueshift to shorter wave lengths occurs when the object is moving towards the observer. The spectral lines for hydrogen are commonly used in cosmology. A dimensionless parameter, Z, given by

$$Z = (\lambda_{observed} - \lambda_{emit}) / \lambda_{emit} \qquad A3$$

provides a measure of the Redshift (where λ is the appropriate wave lengths). Hubble (1929) was one of the first to correlate the Redshift with velocities and distance of remote objects with his cosmological constant. His observations over his lifetime led to the acceptance of the expanding universe hypothesis.

There is another type Red Shift related to photon having energy and thus mass ($E = m c^2$) and affected by gravity which would apply to light rays emitted from dense objects such as black holes.

A Perturbation Theory Model For the Kinetics of the Moon Producing Giant Impact

We wish to develop a kinetics model using Newtonian mechanics to describe the kinetic behavior of the Earth - Sister Planet Giant Impact. We make some assumptions that simplify the model but enable evaluation of the errors in the approximations. We assume that

1.) the Earth and the Sister Planet are in circular orbits. This is an excellent assumption since the eccentricities of Venus, Earth and Mars are 0.007, 0.017 and 0.093, respectively.

2.) The Sister Planet revolves around the Sun in the same orbital plane as Earth. This is also a good assumption since orbital inclinations (with respect to Earth's orbit) of Venus and Mars are 3^0 23' 39" and 1^0 50' 59" , respectively.

3.) The Sister Planet is chosen to be originally located outside Earth's Solar orbit. This is justified since Venus is nearly the size of Earth, Mars is much smaller and there is a deficit in mass density between Earth and Mars. With this choice, the Earth's orbital period will be always shorter than the Sister Planet's so Earth will be passing SP, moving ahead and catching up on the SP's next revolution.

4.) We assume that the mass of SP is on the order of the mass of Mars i.e. about $1/10^{th}$ that of Earth. Other estimates agree with this value.

We select the origin of the Cartesian coordinates frame of reference as the center of Earth. The X axis is the tangent to the Earth's orbit around the Sun with -X being towards the SP as Earth approaches and +X towards the departure of Earth after Earth's passage. We let +Z axis be the distance between the circular orbits. The radius of the circular Earth orbit about the Sun is R such that the radius of the SP's orbit about the Sun is R + Z. We will view the motion from above the North polar axis of Earth such that the Y axis will be downward. In our model then X and Z will be a function of time with no variation in Y since the orbital planes coincide.

Our perturbation model is based on the gravitational force exerted on the SP each time the Earth passes SP inside SP's orbit. We will assume as a first estimate that the SP's orbit is modified with each pass, with Earth's orbit remaining constant at R = 149.5 km from the Sun. The Gravitational Force, Fg, between Earth and SP is given by

$$Fg = G \ Me \ Msp \ / \ Re\text{-}sp^2 \qquad\qquad B1$$

where G = Universal Gravitational Constant = 6.67×10^{-20} km^3 / s^2 kg, Me = mass of Earth = 5.98×1024 kg , Msp = mass of SP = 5.98×10^{23} kg and Re-sp = $(X^2 + Z^2)^{1/2}$ = distance between Earth and SP in km. We shall divide the time dependence of X for the passage of Earth by SP into 4000 incremental intervals in terms of the effect of the gravitational force during the passage. If -Xo is the initial location of the gravitational force calculations and for them to continue until Earth is at +Xo, then the ΔX increments are

$$\Delta X = Xo \ / \ 2000 \qquad\qquad B2$$

The total period, Tsp, for SP is

$$Tsp = 2 \pi (R + Z)^{3/2} / (G Ms)^{1/2} \qquad\qquad B3$$

and the total period for Earth is

$$Te = 2 \pi \; R^{3/2} / (G Ms)^{1/2} \qquad\qquad B4$$

where Ms = mass of Sun. The orbital velocities, Vsp and Ve are thus the orbital distance around the Sun, $2 \pi (R + Z)$ and $2 \pi R$ divided by their periods. The difference in their velocities is

$$\Delta V = Ve \; - Vsp = (G Ms)^{1/2} \{ [1 / R^{1/2}] - [1 / (R + Z)^{1/2}] \} \qquad\qquad B5$$

Thus, the time, Δt, for the Earth to travel the distance ΔX is

$$\Delta t = \Delta X / \Delta V \qquad\qquad B6$$

During this Δt time period the SP experiences an acceleration, a, from Newton's Laws, F = ma,
given by Equation (B1), a = Fg / Msp. This acceleration is manifested by a displacement towards Earth during the time, Δt, by an amount

$$\Delta Re\text{-}sp = (dX2 + dZ2)^{1/2} \; = a \; \Delta t2 / 2 \qquad\qquad B7$$

Figure B1 shows the incremental displacements in the X and Z directions during the time interval, Δt and movement of the SP relative to Earth a distance ΔX. The movement of SP towards Earth and change in Z is given by

$$dZ = [\Delta Re\text{-}sp2 \; - dX2] \, 1/2 \qquad\qquad B8$$

SP has also been pulled ahead in the X direction by an amount
$dX = \Delta Re\text{-}sp \; x \; (X / (X^2 + Z^2)^{1/2}$ so the new X = X + dX for the next interval.
The total displacement of SP through the entire single passage of Earth going by SP is thus

129

$$\text{Total Change in Z each pass} = \square\ \Delta Z\ i \quad i = 1\ \text{to}\ 4000 \qquad\qquad B9$$

The computations are continued for subsequent passes with, for the next pass, the beginning Z being the final Z after the last pass i.e.

$$\text{New Z} = \text{Last initial Z}\ -\ \text{Total Change in Z last pass} \qquad\qquad B10$$

APPENDIX C

Two Independent Variables – Their Covariance and Correlation Coefficients

Covariance is a measure of how two variables (X, Y) change together with respect to an independent variable associated with both. The analysis involves paired data sets, one data point from each independent variable. If the values of Y tend to increase as the values of X increase, the covariance will be positive. If Y tends to decrease as X increases the covariance will be negative. The Pearson's correlation coefficient, r, indicates the strength of the relationship between X and Y. When the absolute value of r is close to 1, there is a strong linear relationship between X and Y. If r is zero (0) there is no correlation. If the value of r is between 0 and 0.3 the correlation is a weak positive relationship. If r is between 0.3 and 0.7 there is a moderate positive relationship. If the value of r is between 0.7 and 1.0 the correlation is a strong relationship.

Then given a set of paired X and Y data $\{(Xi, Yi)\} = 1....n$, the covariance of X and Y is

$$\text{Covariance}(X, Y) = [\, 1 \,/\, (n\text{-}1)\,] \; \Sigma \; (Xi \,-\, \mu x)(Yi \,-\, \mu y) \qquad\qquad C1$$

Where μx and μy are the mean values of each variable set, respectively. To compute the Pearson's correlation coefficient, r, the separate variances of each must be computed i.e.

$$\text{Var }(X) = [\, 1 \,/\, (n\text{-}1)\,] \; \Sigma \; (Xi \,-\, \mu x)^2 \;\text{ and }\; \text{Var }(Y) = [\, 1 \,/\, (n\text{-}1)\,] \; \Sigma \; (Yi \,-\, \mu y)^2 \qquad C2$$

Then

$$r \;=\; \text{Covariance }(X, Y) \,/\, [\, \text{Var X}) \; \text{Var }(Y) \,]^{\,\frac{1}{2}} \qquad\qquad C3$$

REFERENCES

Abad, C. and Vieira, K. Systematic motions in the Galactic plane found in the Hipparcos Catalogue using Herschel's Method. Astron. Astrophys, 442, 745-755, 2005.

Amelin, Y., Krot, A., Hutcheon, I. and Ulyanov, A. Lead isotopic ages of Chrindikes and calcium-aluminum rich inclusions. Science 297, 1678-1683, 2002.

Amelin, Y. A tale of early Earth told in zircon. Science 310, 1914-1915, 2005.

Barry, R. The status of research on glaciers and global glacier recession. A review. Progress in Physical Geology. 30, 285-306, 2005.

Benz, W. Slattery, W. and Cameron, A. The origin of the Moon and the single-impact hypothesis I. Icarus 66, 115-135, 1986.

Benz, W. Slattery, W. and Cameron, A. The origin of the Moon and the single-impact hypothesis II. Icarus 71, 30-45, 1987.

Benz, W., Cameron, A. and Melosh, H. The origin of the Moon and the single-impact hypothesis III. Icarus 81, 113-131, 1989.

Bennett,C. L. et al. First year Wilkinson Microwave Anisotropic Probe observations: Cosmological interpretations. Astrophysics Journal Supplement Ser. 148 (1) 1-27, 2003.

Berger, A., Melice, J. L., and Loutre, M. F, On the origin of the 100-kyr cycles in the astronomical forcing. Paleoceanography. 20, 2005.

Berger, A., Loutre, M. F., and Melice, J. L. Equatorial insolation: from precession harmonics to eccentricity frequencies. Climate of the Past Journal. 2, 131-135, 2006.

Berner, R.A., and Canfield, D.E. A new model for atmospheric oxygen over Phanerozoic time. American Journal of Science. 289, 333-361, 1989.

Berner, R. A., and Kothavala, Z. GEOCARB III: A revised model of atmospheric CO_2 over Phanerozoic time. American Journal of Science. 301, 182-204, 2001.

Bode, H. Network analysis and feedback amplifier design, Princeton, NJ: Van Nostrand. 1945.

Bogojevic, A., Balaz, A. and Belic, A. Spacing of planets in an effective gravitational accretion model. Publ. Astron. Obs. Belgrade. 80, 149-153, 2006.

Bond, G., Showers, W., Cheseby, M., Lotti, R.,Almasi, P., deManocal, P., Priore, P., Cullen, H., Haidas, I., and Bonani, G. A. A pervasive millennial-scale cycle in North Atlantic Holocene and glacial climates. Science 278, 1257-1266, 1997.

Brown, P.T., Li, W., Cordero, E.C., and Mauget, S.A. Global Warming More Moderate Than Worst-Case Models. Duke Environment. April 21, 2015.

Cameron, A. Formation of the pre-lunar accretion disk. Icarus 56, 195-201, 1985.

Cameron, A. and Ward, W. The origin of the Moon. Lunar Science VII, 120-122, 1976.

Cameron, A. and Benz, W. The origin of the Moon and the single-impact hypothesis IV. Icarus 92, 204-216, 1991.

Canup, R and P. Esposito. Accretion of the Moon from an impact-generated disk. Science Digest 119, 427-446.1996.

Canup, R. New model reconciles the moon's Earth-like composition with the giant impact theory of formation. Science News. October 17, 2012.

Canup, R. Forming a Moon with an Earth-like composition via a giant impact. Science 338, 1052-1055, 2012.

Cerling, T.E., Wang, Y., and Quade, J. Expansion of C4 ecosystems as an indicator of global ecological change in the late Miocene. Nature 361, 344-345, 1993.

Chambers, J. Planetary accretion in the inner Solar system. Earth and Planetary Science Letters. 223, 241-252. 2004.

Cuk, M. and Stewart, S. Making the Moon from a fast-spinning Earth: A giant impact followed by resonant despinning. Science 338. 1047-1052. 2012.

Dahl-Jensen, D. et al. Eemian interglacial reconstructed from a Greenland folded ice core. Nature 493, 489-494, 2013.

Dansgaard, W. The isotopic composition of natural waters with special reference to the Greenland ice cap. Meddelelser Om Greenland, Bd. 165, Nr. 2. 1961.

Ekart, D. D., Cerling, T.E., Montanez, I.P., and Tabor, N.J. A 400 million year carbon isotope record of pedogenic carbonate: Implications for atmospheric carbon dioxide. American Journal of Science. 299, 805-817, 1999.

Falkowski, P., Scholes, R. J., Boyle, E., Canadell, J., Canfield, D., Elser, J., Gruber, N., Hibbard, K., Hogberg, P., Linder, S., Mackenzie, F. T., Moore III, B., Pedersen, T., Rosenthal, Y., Seltzinger, S., Smetacek, V., and Steffen, W. The global carbon cycle: A test of our knowledge of Earth as a system. Science 290, 291-296, 2000.

Fischer, D. and Valenti, J. Stars rich in heavy metals tend to harbor planets. UCBerkley News, 21 July 2003.

Fletcher, B. J., Brentnall, S. J., Quick, P. W., and Beerling, D. J. BRYOCARB: A process-based model of thallose liverwort carbon dioxide isotope fractionation in response to CO_2, O_2, light and temperature. Geochimica et cosmochimica. 70, 5676-5691, 2006.

Gore, A. Earth in the balance. Rodale Publishers, Inc. New York. 2005.

Gore, A. An inconvenient truth. Rodale Publishers, Inc. New York. 2006.

Gott, J. Where did the Moon come from? Astronomical Review, May 17, 2011.

Hansen, J., Sato, M., Kharecha, P., Russwell G., Lea, D. W., and Siddal, M. Climate change and trace gasses. Philosophical Transactions of the Royal Society A. 365, 1925-1954, 2007.

Hansen, J. The threat to the planet: How can we avoid dangerous human-made climate change. Remarks made on 21 November 2006 on acceptance of WWF Duke of Edinburgh Conservation Medal at St. James Palace, London. 2006.

Harms, A., Schopfi, K., Miley, G. and Kingham, D. Principles of fusion energy. Amazon Publishers, 2000.

Hartmann, W. and Davis, D. Satellite-sized planetesimals and lunar origin. Icarus 24, 504-515, 1975.

Hieb, M. Plant fossils of West Virginia – Climate and the Carboniferous Period. Internet – 2006.

Hubble, E. A relationship between distances and radial velocity among extra-galactic nebulae. Proceedings of the National Academy of Sciences. 15, 168-173, 1929.

Ida, S., Canup, R. and Stewart, G. Lunar origin from an impact-generated disk. Nature Letters. 389, 353-357. 1997.

Indermuhle, A., Monnin, E., Stauffer, B., and Stocker, T. F. Atmospheric CO_2 concentration from 60 to 20 kyr BP from the Taylor Dome ice core, Antarctica. Geophysical Research Letters, 27, 735-738, 2000.

IPCC, Climate change: The IPCC assessment, Eds. Houghton, J.T., Jenkins, G.J., and Ephraums, J.J. Cambridge Press, Cambridge, U. K. 1990.

IPCC, Climate change 2007: Synthesis report. IPCC Secretariat, World Meteorological Organization, Geneva, Switzerland, 2007.

IPCC, Climate Change 2013. Working Group I Contribution to the Fifth Assessment Report (AR5), Stockholm, 23-26 September 2013.

Jouzel, J. et al (31 others). EPICA dome C ice core 800KYr deuterium data and temperature estimates. Science, 317, 793-797, 2007.

Kirby, M., Jones, D. and MacFadden, B. Lower Miocene stratigraphy along the Panama Canal and its bearing on the Central American peninsula. PLOS One, July 30, 2008.

Kirschvink, J. L., and Kopp, R. E. Arguments for the late evolution of oxygenic photosynthesis at ~ 2.3 Ga: A trigger for the Paleoproterozoic snowball Earth. Geophysical Research Abstracts. 7, 11197, 2005.

Knie, K. et al. 60Fe anomaly in a deep –sea manganese crust and implications for a near-by supernova source. Physical Review Letters 93, 171103, 2004.

Kobashi, T., Severinghaus, J.P., and Barnola, J-M. 4 ± 1.5 °C abrupt warming 11,270 yr ago identified from trapped air in Greenland ice. Earth and Planetary Science Letters 268, 397-407, 2008.

Kopp, R.E., Kirschvink, J.L., Hilburn, I.A., and Nash, C.Z. The paleoproterozoic snowball Earth: A climate disaster triggered by the evolution of oxygenic photosynthesis. Proceedings of the National Academy of Science, 102, 11131-11136, 2005.

Koppes, S. Titanium paternity test fingers Earth as moon's only parent. University of Chicago News, March 28, 2012.

Krot, A.,Meibom, K., Weisberg, M and Keil, K. The CR chondrite clan: Implications for early solar system processes. Meteor. And Planet . Sci. 37, 1451-1490, 2002.

Kroupa, P. The initial mass function of stars: evidence of uniformity in variable systems. Science 295, 82-91, 2002.

Laskar, J., Robutel, P., Joutel, F., Gastineau, M.,Correia, A. and Levard, B. A long-term numerical solution for the insolation quantities of the Earth. Astronomy and Astrophysics. 428,261-285, 2004.

Leonard, B. Human lung cancer risks from Radon – Influence from bystander and adaptive response non-linear dose response effects. Amazon Publishers. 2012.

Leutwyler, K. Rise of the Himalayas may have started the monsoons and ice age. Scientific American Journal, May 3, 2001.

Lynch-Stieglitz, J., Curry, W. and Stowey, N. Weaker Gulf Stream in the Florida Straits during the last glacial maximum. Letters to Nature 402, December 9, 1999.

Lisiecki, L. E. and Raymo, M. E. A Pliocene-Pleistocene stack of 57 globally distributed benthic 18O records. Paleoceanography 20: 1003, 2005.

Loutre, M. F., and Berger, A. Future changes: are we entering an exceptionally long interglacial? Climate Change 46, 61-90, 2000.

Lowe, D. R., and Tice, M. M. Geological evidence for Archean atmosphere and climate evolution: Fluctuating levels of CO2, CH4 and O2 with an overriding tectonic control. Geology, 32, 493-496, 2004.

Luthi, M., Le Floch, M., Bereiter, B., Blunier, T.,Barnola, J.M., Siegenthaler, U., Raynaud, D., Jouzel, J., Fischer,H., Kawamura, K., and Stocker, T.F. EPICA dome C ice core 800Kyr carbon dioxide data. Nature 453, 379-382, 2008.

Mann, M., Bradley, R., Hughes, M. Northern hemisphere temperatures during the past millennium: Inferences, uncertainties, and limitations. Geophysical Research Letters 26, 759, 1999.

McIntyre, S. and McKitrick, R. Hockey sticks, principal components, and spurious significance. Geophysical Research Letters. 32, L03710, 2005.

McManus, J., Oppo, D., Cullen, J. and Healey, S. Marine isotope stage 11 (MIS 11): Analog for Holocene and future climate? Geophysical Monograph Series, 137, 69-85, 2003.

Melosh, H. J., and Vickery, A. M. Impact erosion of the primordial atmosphere of Mars. Nature, 338, 487-489, 1989.

Milankovitch, M. (English translation) Canon of insolation and the ice age problem. With introduction and biographical essay by Nikola Pantic. 636 pages. ALVEN Global Publishers. 1998.

Miller, K.G., Wright, J.D., and Browning J.V. Visions of ice sheets in a greenhouse world. In: Ocean Chemistry Throughout the Phanerozoic, Payton, A., and De La Rocha. Special Issue. 217, 323-338, 2005.Marine Geology. 2005.

Minkel, J. R. Appalachians triggered ancient ice age. Scientific American Journal, October 25, 2006.

Minkel, J. R. Giant asteroid flattened half of Mars, studies suggest. Scientific American Journal, June 25, 2008.

Muller, R. A., and MacDonald, G. J. Spectrum of 100-kyr glacial cycle: Orbital inclination, not eccentricity. Proceeding of the National Academy of Science. 94, 8320-8334, 1997.

Navarro-Gonzalez, R., McKay, C.P., Nina Mvondo, D., Coli, P., and Raulin, F. Nitrogen fixation rate by lightning during the evolution of the atmosphere. Geophysical Research Abstracts, 5, 3277, 2003.

Ninkovic, S. and Trajkovska, V. On the mass distributions of stars in the Solar neighborhood. Serbian Astronomical Journal. 172, 1

Ohmoto, H., Watanabe, Y., and Kumazawa, K. Evidence from massive siderite beds for a CO2 – rich atmosphere before ~ 1.8 billion years ago. Nature 429, 395-399, 2004.

Olson, J.M. Photosynthesis in the Archean Era. Photosynthesis Research. 88, 109-117, 2006.

Osborne, C.P., and Beerling, D.J. Review. Natures green revolution: the remarkable evolutionary rise of C 4 plants. Science 361, 173-194, 2005.

Pagani, M., Zachos, J. C., Freeman, K. H., Tipple, B., and Bohaty, S. Paleogene atmospheric carbon dioxide reconstruction. NOAA Satellite and Information Service. U. S. Department of Commerce, Boulder, Colorado 2005.

Pagani, M., Caldeira, K. and Archer, D. An ancient carbon mystery. Science 314, 1556-1557. 2006.

Paniello, R., Day, J. and Moynier, F. Zinc isotope evidence for the origin of the moon. Nature Letters. 490, 375-379. 2012.

Petit, J. Jouzel, J. et al Climate and atmospheric history of the past 420,000 years from the Vostok ice core, Antarctica. Nature 399, 429-436, 1999.

Rahmstorf, S. Ocean circulation and climate during the past 120,000 years. Nature 419, 207-214, 2002.

Raymo, M.E., and Nisancioglu, K. The 41 kyr world: Milankovitch's other unsolved mystery. Paloceanography, 18, DOI 10.1029/2002PA000791, 2003.

Raymo, M.E., Oppo, D.W., Flower, B.P., Hodell, D.A., McManus, J., Venz, K.A., Kleiven, K.F. and MaIntyre, K. 2004. Stability of North Atlantic water masses in face of pronounced natural climate variability. Paleoceangraphy. 19. 421-436.

Raymo. M. E. and Mitrovica, J. X. Collapse of polar ice sheets during the stage 11 interglacial. Nature Letters, 10891, 2012.

Reynolds, S., Borkowski, K. et al. The youngest galactic supernova remnant: G1.9+0.3. Astrophysical Journal Letters 680.L41-L44. 2008.

Rial, J. et al. Nonlinearities, feedback and critical thresholds within the Earth's climate system. CiteSeer, 2004

Rohling, E., Braun, K., Grant, K., Kucera, M., Roberts, A., Siddall, M. and Trommer, G. Comparison between Holocene and Marine Isotope Stage-11 sea level histories. Earth and Planetary Science Letters. 291,97-105, 2010.

Planetary Change 41, 95-109, 2004.

Salpeter, E. E. The luminosity function and stellar evolution. Astrophysical Journal, 121, 161, 1955.

Savage, J. and Papworth, D. Constructing a 2B calibration curve for retrospective dose reconstruction. Radiation Protection Dosimetry. 88, 69-76, 2000.

Scotese, C. R. PaleoMap. Internet, 2003.

Siddall, M., Rohling, E.J., Almogi-Labin, C., Hemleben, D., Meischner, I., Schmelzer, , and Smeed, D.A. Sea-level fluctuations during the last glacial cycle. Nature 423, 853-858, 2003.

Siegenthaler, U. et al. Stable carbon dioxide cycle – Climate relationship during the late Pleistocene. Science 310, 1313, 2005.

Singh, B. P., and Lee, Y. I. Atmospheric pCO2 and climate during late Eocene (36 ± 5 Ma) on the Indian subcontinent. Current Science, 92, 518-520, 2007.

Soon, W. Quantitative implications of the secondary role of carbon dioxide climate forcing in the past glacial-interglacial cycles for the likely future climate impacts and anthropogenic greenhouse-gas forcings. Physical Geography, 2007.

Stott, L., Timmermann, A., and Thunell, R. Southern hemisphere and deep-sea warming led deglacial atmospheric CO2 rise and tropical warming. Science. 318, 435-436, 2007.

Souchez, R., Janssens, L., Lemmens, M., and Stauffer, B. Very low oxygen concentration in basal ice from Summit, Central Greenland. Geophysical Research Letters, 22, 2001-2004, 1995.

Spitzer, R. J. New Proofs for the Existence of God. William B. Eerdmans Publishing Company, Cambridge, U. K. 2010.

Symonds, R. B., Rose, W. I., Bluth, G., and Gerlach, T. M. Volcanic gas studies: methods, results, and applications. In: Carroll, M. R., and Holloway, J. R. Eds. Volatiles in Magnas: Mineralogical Society of America, Reviews in Mineralogy, 30, 1-66, 1994.

Tyrrell, T., Shepherd, J.G., and Castle, S. The long-term legacy of fossil fuels. Tellus B. 59,564-572, 2007.

Tziperman, E., Raymo, M.E., Huybers, P., and Wunsch, C. Consequences of pacing the Pleistocene 100 kyr ice ages by nonlinear phase locking to Milankovitch forcing. Paleocean0graphy, 21, PA4206, 1-11, 2006.

Walton, J. C. The chemical composition of the Earth's original atmosphere. Origins, 3, 66-84, 1976.Wiechert, U. et al. Oxygen isotopes and the Moon-forming giant impact. Science 294, 345-348, 2001.

Wilson, R. C., and Mordvinov, A. V. Secular total solar irradiation trend during Solar cycles 21-23. Geophysical Research Letters, 30,3.1-3.4, 2003.

Wrigley, T.M.L., and Kelly, P.M. Holocene climate change, 14C wiggles and variation in Solar irradiance. Philosophical Transactions of the Royal Society of London. Series A 330, 547-560, 1990.

Yang, M., and Rial, J.A. Internal oscillations of the thermohaline circulation and the Dansgaard-Oeschger events of the last ice age. American Geophysical Union, December 2006.

INDEX

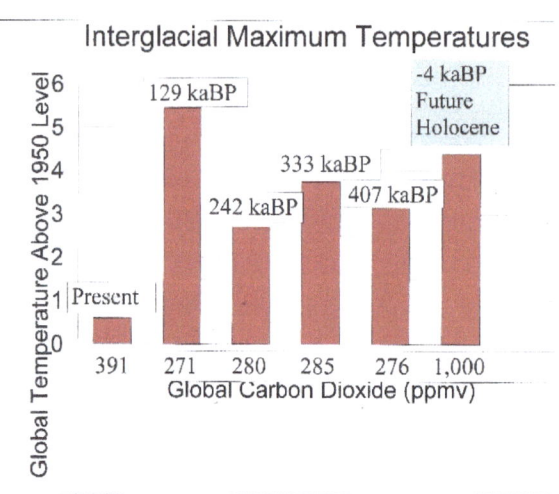

Figure 62, Maximum Global Surface Air Temperatures for Past Four and Present Interglacial Periods, EPICA data.

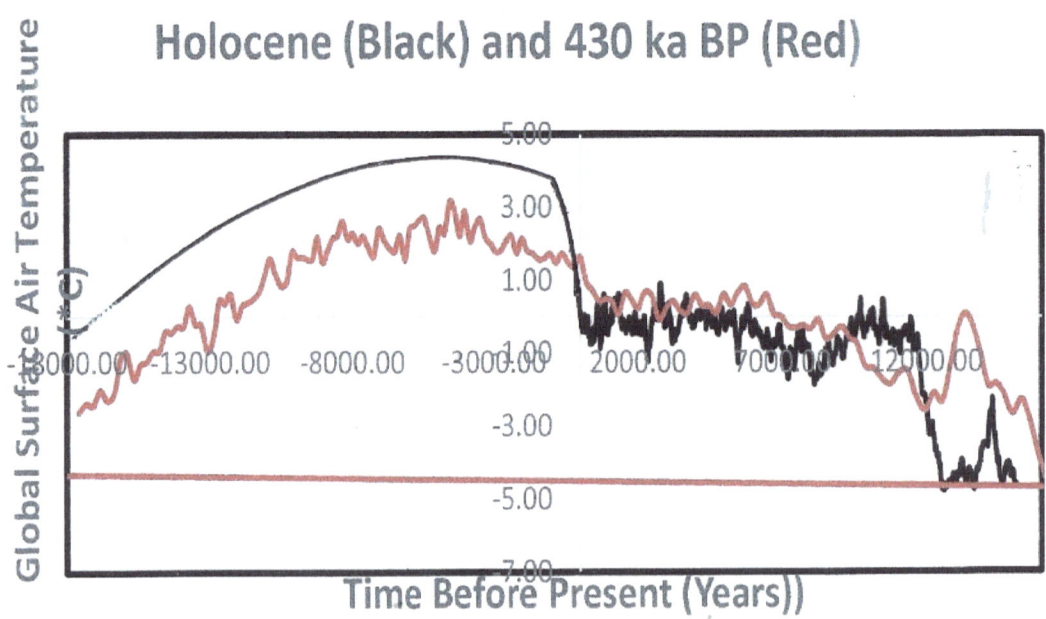

Figure 92, Estimated Global Surface Air Temperature Through the Remainder of Our Present Holocene Interglacial Period

Bobby Leonard, PhD Engineering

Arthur Lucas, PhD Physics